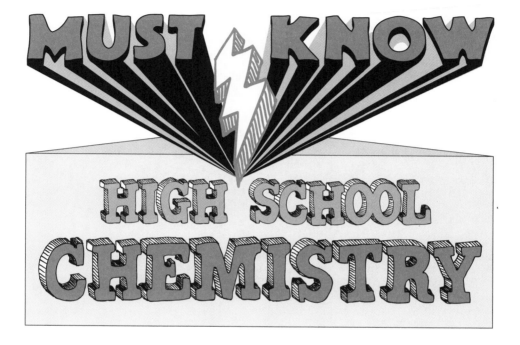

MUST KNOW

HIGH SCHOOL CHEMISTRY

Second Edition

John T. Moore, EdD

Richard Langley, PhD

Mc Graw Hill

New York   Chicago   San Francisco   Athens   London   Madrid
Mexico City   Milan   New Delhi   Singapore   Sydney   Toronto

1 2 3 4 5 6 7 8 9   LCR   27 26 25 24 23 22

ISBN   978-1-264-28617-1
MHID   1-264-28617-1

e-ISBN   978-1-264-28618-8
e-MHID   1-264-28618-X

Interior design by Steve Straus of Think Book Works.
Cover and letter art by Kate Rutter.

McGraw Hill books are available at special quantity discounts to use as premiums and sales promotions or for use in corporate training programs. To contact a representative, please visit the Contact Us pages at www.mhprofessional.com.

McGraw Hill is committed to making our products accessible to all learners. To learn more about the available support and accommodations we offer, please contact us at accessibility@mheducation.com. We also participate in the Access Text (www.accesstext.org), and ATN members may submit requests through ATN.

Contents

16 Buffers and Additional Equilibria   283

17 Entropy and Free Energy   305

18 Electrochemistry   319

19 Chemistry of the Elements   341

Introduction

Welcome to your new chemistry book! Let us explain why we believe you've made the right choice with this new edition. This probably isn't your first go-round with either a textbook or other kind of guide to a school subject. You've probably had your fill of books asking you to memorize lots of terms. This book isn't going to do that—although you're welcome to memorize anything you take an interest in. You may also have found that a lot of books make a lot of promises about all the things you'll be able to accomplish by the time you reach the end of a given chapter. In the process, those books can make you feel as though you missed out on the building blocks that you actually need to master those goals.

With *Must Know High School Chemistry,* we've taken a different approach. When you start a new chapter, right off the bat you will immediately see one or more **must know** ideas. These are the essential concepts behind what you are going to study, and they will form the foundation of what you will learn throughout the chapter. With these **must know** ideas, you will have what you need to hold it together as you study, and they will be your guide as you make your way through each chapter.

To build on this foundation, you will find easy-to-follow discussions of the topic at hand, accompanied by comprehensive examples that show you how to apply what you're learning to solving typical chemistry questions. Each chapter ends with review questions—more than 300 throughout the book—designed to instill confidence as you practice your new skills.

This book has other features that will help you on this chemistry journey of yours. It has a number of sidebars that will both provide information or just serve as a quick break from your studies. The **BTW** sidebars ("by the way") point out important information as well as study tips and exceptions to the rule. Every once in a while, an 🌐 **IRL** sidebar ("in real life") will tell you what you're studying has to do with the real world; other IRLs may just be interesting factoids.

But that's not all—this new edition has taken it a step further. We know our Chemistry students well, and we want to make sure you're getting the most out of this book. We added new **EASY MISTAKE** sidebars that point out common mistakes and things *not* to do. For those needing a little assistance, we have our **EXTRA HELP** **A+** feature, where more challenging concepts, topics, or questions are given some more explanation. And finally, one special note for the teachers (because we didn't forget about you!)—a **Teacher's Guide** section at the back of the book is a place where you can go to find tips and strategies on teaching the material in the book, a behind-the-scenes look at what the authors were thinking when creating the material, and resources curated specifically to make your life easier!

In addition, this book is accompanied by a flashcard app that will give you the ability to test yourself at any time. The app includes more than 100 "flashcards" with a review question on one "side" and the answer on the other. You can either work through the flashcards by themselves or use them alongside the book. To find out where to get the app and how to use it, go to the next section, The Flashcard App.

Before you get started, though, let me introduce you to your guides throughout this book. John and Rich—er, Drs. Moore and Langley—have taught chemistry to students throughout their university careers. They have a clear idea about what you should get out of a chemistry course and have developed strategies to help you get there. They also have seen the kinds of trouble that students can run into, and they are experienced hands at solving those difficulties. In this book, they apply that experience both to showing you the most effective way to learn a given concept as well as how to extricate yourself from traps you may have fallen into. They will be trustworthy guides as you expand your chemistry knowledge and develop new skills.

Before we leave you to Drs. Moore and Langley's surefooted guidance, let us give you one piece of advice. While we know that saying something "is the *worst*" is a cliché, if anything *is* the worst, it's the mole concept. Let Drs. Moore and Langley introduce you to the concept and show you how to apply it confidently to your chemistry work. Take our word for it, mastering the mole concept will leave you in good stead for the rest of your chemistry career.

Good luck with your studies!

The Editors at McGraw Hill

The Flashcard App

This book features a bonus flashcard app. It will help you test yourself on what you've learned as you make your way through the book (or in and out). It includes 100-plus "flashcards," both "front" and "back." It gives you two options as to how to use it. You can jump right into the app and start from any point that you want. Or you can take advantage of the handy QR codes near the end of each chapter in the book; they will take you directly to the flashcards related to what you're studying at the moment.

To take advantage of this bonus feature, follow these easy steps:

Search for **McGraw Hill Must Know** App from either Google Play or the App Store.

↓

Download the app to your smartphone or tablet.

↓

Once you've got the app, you can use it in either of two ways.

↙ ↘

| Just open the app and you're ready to go. | Use your phone's QR code reader to scan any of the book's QR codes. |
| You can start at the beginning, or select any of the chapters listed. | You'll be taken directly to the flashcards that match your chapter of choice. |

↘ ↙

Be ready to test your chemistry knowledge!

1 Getting Started in Chemical Calculations

MUST KNOW

- The unit conversion method and the SI system are essential to chemistry calculations.

C hemistry is full of calculations. We want you to develop the knowledge and strategies needed to solve these problems. In this chapter, we will review the International System of Units (SI system), significant figures, and basic problem-solving techniques, such as the unit conversion method. This problem-solving method is many times called by different names, such as the factor-label method or dimensional analysis. Finally, be familiar with the operation of your calculator. (A scientific calculator will be the best for chemistry purposes.) Be sure that you can correctly enter a number in scientific notation. It would also help if you set your calculator to display in scientific notation. Refer to your calculator's manual for information about your specific brand and model. Chemistry is not a spectator sport, so we will need to work a lot of problems!

How to Study Chemistry

The mastery of chemistry, like most sciences, involves active participation by the one studying it. We suggest reading the material in depth and working problems. This book is designed to help you grasp the basic concepts and to help you learn how to work the problems associated with the material. These specific tips will help you in your study of chemistry:

- Strive for understanding, not just memorization.

- Study some chemistry every day—long study sessions are not nearly as effective as shorter, regular study sessions that usually take less overall time.

- Work many, many problems, but again strive for understanding—it is a waste of time to simply memorize how to do a problem; and it is also a waste of time to simply look over the solution for a problem without striving to understand why a certain procedure was followed. You can prove that you understand the problem by working it without looking at the solution.

■ Nomenclature, the naming of chemical compounds, is extremely important. When the time comes, learn the rules and apply them. Calling a chemical compound by the wrong name is certainly not the way to impress anyone knowledgeable.

You will be doing many problems in your study of chemistry. Here are some specific suggestions to help you in your problem solving:

■ Identify and write down what quantity you wish to find.

■ Extract and write down just the pertinent information from the problem, especially the numbers *and* units—this is especially important for long word problems.

■ Identify and write down any equations or relationships that might be useful.

■ Look for relationships among the information from the problem, the equations, and the quantity you wish to find.

■ Use the unit conversion method (explained later in this chapter) in solving for the desired quantity. You may be tempted to take shortcuts on some of the problems in this chapter, but you will pay for this later when you deal with more involved calculations.

The Macroscopic and Microscopic Levels of Chemistry

You can view many things in chemistry on both the macroscopic level (the level that we can directly observe) and the microscopic level (the level of atoms and molecules). Many times, observations at the macroscopic level can influence the theories and models at the microscopic level. Theories and models at the microscopic level can suggest possible experiments at the macroscopic level. We express the properties of matter in both ways.

Matter (anything that has mass and occupies space) can normally exist in one of three states: solid, liquid, or gas. At the macroscopic level, a **solid** has both a definite shape and a definite volume. At the microscopic level, the particles that make up a solid are very close together and are restricted to a very regular framework called a crystal lattice. Molecular motion (vibrations) exists, but it is slight.

Macroscopically, a **liquid** has a definite volume but no definite shape. It conforms to the shape of its container. Microscopically, the particles are still very close together but moving much more than in the solid.

A **gas**, at the macroscopic level, has neither a definite shape nor volume. It expands to fill its container. The microscopic view is that the particles are far apart, moving rapidly with respect to each other, and act independently of each other.

We indicate the state of matter that a substance is in by (s), (l), or (g). Thus, $H_2O(g)$ would represent gaseous water (steam), $H_2O(l)$ would represent liquid water, while $H_2O(s)$ would represent solid water (ice).

Units of Measurement (SI System)

The measurement system that you will most likely encounter is the SI (revised metric) system. Each quantity (such as mass and volume) has a base unit and sometimes a prefix that modifies the base unit. The prefixes are the same for all quantities and are based on a decimal system. The following are some basic units; we will introduce others as needed in later chapters:

length	meter	m
mass	kilogram	kg
volume	cubic meter	m^3 (SI)
	or liter	L (metric)
temperature	Kelvin	K

Some of the prefixes that we will be using in the SI system are in the following table. You will want to be familiar with these and others.

Prefix	Abbreviation	Meaning
micro-	μ	0.000001 or 10^{-6}
milli-	m	0.001 or 10^{-3}
centi-	c	0.01 or 10^{-2}
deci-	d	0.1 or 10^{-1}
kilo-	k	1,000 or 10^{3}
mega-	M	1,000,000 or 10^{6}

Sometimes it is necessary to convert from a measurement in the English system to a measurement in the SI system. (The English system is sometimes referred to as the US customary system of units or the English system of units.) There are numerous SI/English conversions. We will be using the following in many of our examples:

length   1 inch (in)   = 2.54 centimeters (cm)

mass   1 pound (lb)   = 453.59 grams (g)

volume   1.057 quart (qt) = 1 liter (L)

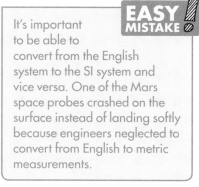

EASY MISTAKE

It's important to be able to convert from the English system to the SI system and vice versa. One of the Mars space probes crashed on the surface instead of landing softly because engineers neglected to convert from English to metric measurements.

Significant Figures

We will be dealing with two types of numbers in chemistry—exact and measured numbers. Exact values have no uncertainty associated with them. They are exact normally by definition. There are exactly 12 items in a dozen, 36 inches in a yard, and so forth. Most conversions, especially SI-to-SI conversions and US-to-US conversions, involve exact numbers. Measured values, like the ones you will be dealing with in lab, have uncertainty associated with them because of the limitations of our measuring instruments. When those measured values appear in calculations, the answer must reflect that combined uncertainty by the number of significant figures that you report in the final answer. The more significant figures reported, the greater the certainty in the answer.

The measurements that you use in calculations may contain varying numbers of significant figures, so carry as many as possible until the end and then round off the final answer. The least precise measurement will determine the significant figures reported in the final answer. Determine the number of significant figures in each measured value (not the exact ones), and then, depending on the mathematical operations involved, round off the final answer to the correct number of significant figures. Here are the rules for determining the number of significant figures in a measured value:

- All nonzero digits (1, 2, 3, 4, and so on) are significant.

- Zeroes between nonzero digits are significant.

- Zeroes to the left of the first nonzero digit are not significant.

- Zeroes to the right of the last nonzero digit are significant if there is a decimal point present, but not significant if there is no decimal point.

The last rule is a convention that we will be using, but some sources may use alternative methods. By these rules, 0.0320400 would contain six significant figures, but 320,400 would only contain four.

Another way to determine the number of significant figures in a number is to express it in scientific (exponential) notation. The number of digits shown is the number of significant figures. For example, 2.305×10^{-5} would contain four significant figures.

In determining the number of significant figures you will express in the final answer, the following rules apply:

■ For addition and subtraction problems, round the answer off to the same number of decimal points as the measurement with the fewest decimal places.

18.256 cm + 7.25 cm + 2.7 cm = 28.206 cm = 28.2 cm
(rounded to the tenths place because of the 2.7)

■ For multiplication and division problems, round off the answer to the same number of significant figures in the measurement with the fewest significant figures.

7.253 m × 3.52 m = 25.53056 m^2 = 25.5 m^2
(rounded to three significant figures because of the 3.52)

In this book, we will tend to be very strict in rounding off the final answer to the correct number of significant figures.

Carry as many numbers as possible throughout the calculation and only round off the final answer.

Problem Solving by the Unit Conversion Method

In this section, we will introduce one of the most common methods for solving problems. This is the unit conversion method. It will be very important for you to take time to make sure you fully understand this method. You may need to review this section from time to time. The unit conversion method, sometimes called the factor-label method or Dimensional Analysis, is a method for simplifying chemistry problems.

This method uses units to help you solve the problem. While slow initially, with practice it will become much faster and second nature to you. If you use this method correctly, it is nearly impossible to get the wrong answer. For practice, you should apply this method as often as possible, even though there may be alternatives. Eventually, you will not need to write down every step on paper.

Let's use the question of "How many feet are in 3.5 miles?" to illustrate how to apply the unit conversion method. First, we will organize the information by writing the given value (3.5 mi) and the unit for the answer, separated by an equal sign.

$$3.5 \text{ mi} = ? \text{ ft}$$

We now need a relationship involving miles. It does not matter what relationship we use, any one will help you find the answer. However, the ideal relationship would involve the unit you now have and the unit you are seeking. In this case, the relationship "5280 ft = 1 mi" is ideal. This relationship could appear two different ways in a problem. These two ways are:

$$\left(\frac{5280 \text{ ft}}{1 \text{ mi}} \right) \text{ or } \left(\frac{1 \text{ mi}}{5280 \text{ ft}} \right)$$

In this situation, we will want to use the first of these two relationships. We know that we need this one to cancel the initial unit of mile with the unit of mile in the denominator of this fraction. (If we were to use the other term, we would get mi^2/ft, which is not only the wrong unit but one of limited use.) Now, we combine this relationship with the initial value:

$$3.5 \text{ mi} \left(\frac{5280 \text{ ft}}{1 \text{ mi}} \right) = ? \text{ ft}$$

▸ This could also be written as:

$$\left(\frac{3.5 \text{ mi}}{1}\right) \times \left(\frac{5280 \text{ ft}}{1 \text{ mi}}\right) = ? \text{ ft}$$

▸ We can now cancel identical units:

$$3.5 \; \cancel{\text{mi}} \left(\frac{5280 \text{ ft}}{1 \; \cancel{\text{mi}}}\right) = ? \text{ ft}$$

▸ The only units remaining on the left are the ones we are seeking (feet). Now that we have the proper units, we can enter the values into a calculator to finish the problem:

$$3.5 \text{ mi} \left(\frac{5280 \text{ ft}}{1 \text{ mi}}\right) = 18{,}480 \text{ ft} = 18{,}000 \text{ ft}$$

(proper number of significant figures)

▸ or

$$3.5 \text{ mi} \left(\frac{5280 \text{ ft}}{1 \text{ ft}}\right) = 18{,}480 \text{ ft} = 1.8 \times 10^4 \text{ ft}$$

Unit conversion problems can appear to be more complicated than this one. However, they are not. They just involve more simple steps.

EXAMPLE

▸ Suppose the original question was "How many inches are in 3.5 miles?" We could use our answer of the number of feet in 3.5 mi and apply the relationship "12 in = 1 ft" to do one more conversion:

$$18{,}000 \text{ ft} \left(\frac{12 \text{ in}}{1 \text{ ft}}\right) = ? \text{ in}$$

▶ This step is just like the original conversion of miles to feet. We can now enter these values into a calculator:

$$18,000 \text{ ft} \left(\frac{12 \text{ in}}{1 \text{ ft}} \right) = 216,000 \text{ in} = 220,000 \text{ in (or } 2.2 \times 10^5 \text{ in)}$$

▶ In problems such as the mile-to-inch conversion, it is simpler to combine the two calculations into one, which avoids intermediate rounding (the 18,000 ft is a rounded number). In this case, we write the first step:

$$3.5 \text{ mi} \left(\frac{5280 \text{ ft}}{1 \text{ mi}} \right) = ? \text{ in}$$

▶ Now, insert the second step and calculate the answer:

$$3.5 \text{ mi} \left(\frac{5280 \text{ ft}}{1 \text{ mi}} \right)\left(\frac{12 \text{ in}}{1 \text{ ft}} \right) = 220,000 \text{ in (or } 2.2 \times 10^5 \text{ in)}$$

▶ This method lets us get the answer without calculating an intermediate value. It also avoids the possibility of too much rounding, which is always a concern in multistep problems.

▶ If you did not know the miles-to-feet relationship, you might do the problem as follows:

$$3.5 \text{ mi} \left(\frac{1760 \text{ yd}}{1 \text{ mi}} \right)\left(\frac{36 \text{ in}}{1 \text{ yd}} \right) = 2.2 \times 10^5 \text{ in}$$

BTW

Only round your final answer to the proper number of significant figures.

▶ There are numerous ways to do a problem correctly. Just use the relationships you already know or you learn as you work through this book.

You can work all unit conversion problems by this procedure. In some cases, such as the miles-to-feet problem, only one step is necessary. In other cases, such as the miles-to-inches problem, more than one step is necessary.

However, all steps are of the same type. You will cancel one unit and get a new unit. This process continues until the new unit matches the one that you are seeking.

One of the concepts students often see in the introductory material of chemistry texts is density. Density is defined as the mass per unit volume (density = mass/volume). We treat this relationship as a simple definition and not just a mathematical equation for you to memorize. We will now explore how to use this definition in the unit conversion method.

EXAMPLE

▶ In this example, we wish to find the density, in g/cm^3, of a sample of a liquid, given that 2.00 ft^3 of the liquid weighs 97.5 lb.

▶ Let's begin by isolating the information from the problem:

volume = 2.00 ft^3 mass = 97.5 lb density = ? g/cm^3
(making a list is a good habit)

▶ (You will notice that, in this problem, we do not concern ourselves with the distinction between weight and mass.)

▶ Once we have isolated the information, we now write the definition of density:

$$\text{density} = \frac{(\text{mass})}{(\text{volume})}$$

▶ According to the unit conversion method, we may begin with any mass unit over any volume unit and have a density. Our isolated information supplies a mass and a volume. We can now enter this information into the definition:

$$\text{density} = \frac{(\text{mass})}{(\text{volume})} = \frac{(97.5 \text{ lbs})}{(2.00 \text{ ft}^3)}$$

▶ This gives us a density in terms of lb/ft^3. While this is a density, it does not have the desired units. To complete the problem, we need to make two conversions: pounds to grams and cubic feet to cubic centimeters. We can do these conversions separately, and in any order. There are several hundred ways to finish this problem correctly. Everyone doing the problem will develop their own "correct" method. The correct method for you will be the one that uses the conversions you know. For example, here are some ways of changing 97.5 lb to grams:

$$(97.5 \text{ lb})\left(\frac{453.59 \text{ g}}{1 \text{ lb}}\right) = 4.42 \times 10^4 \text{ g}$$

$$(97.5 \text{ lb})\left(\frac{1 \text{ kg}}{2.205 \text{ lb}}\right)\left(\frac{10^3 \text{ g}}{1 \text{ kg}}\right) = 4.42 \times 10^4 \text{ g}$$

$$(97.5 \text{ lb})\left(\frac{16 \text{ oz}}{1 \text{ lb}}\right)\left(\frac{28.349 \text{ g}}{1 \text{ oz}}\right) = 4.42 \times 10^4 \text{ g}$$

▶ Clearly any of these procedures, and many more, will work. In this case, we will use the first:

$$\text{density} = \frac{(\text{mass})}{(\text{volume})} = \frac{(97.5 \text{ lb})}{(2.00 \text{ ft}^3)}\left(\frac{453.59 \text{ g}}{1 \text{ lb}}\right)$$

▶ Now that the mass conversion is complete, we can move on to a volume conversion. As with the mass conversion, there are many different correct volume conversions. In this case, the conversion will relate, in some way, to the length unit of feet. Thus, we might use 12 in = 1 ft. If we incorporate this, we get:

$$(2.00 \text{ ft}^3)\left(\frac{12 \text{ in}}{1 \text{ ft}}\right)$$

▶ This will leave us with (ft^2 · in). This is not very useful. We need to remember that ft^3 means ft × ft × ft. We can use the following to cancel the ft^3:

$$(2.00 \text{ ft}^3)\left(\frac{12 \text{ in}}{1 \text{ ft}}\right)\left(\frac{12 \text{ in}}{1 \text{ ft}}\right)\left(\frac{12 \text{ in}}{1 \text{ ft}}\right) = \dots \text{ in}^3$$

▶ An easier way of writing this would be:

$$\left(2.00 \text{ ft}^3\right)\left(\frac{12 \text{ in}}{1 \text{ ft}}\right)^3 = 3456 \text{ in}^3 \text{ (unrounded)}$$

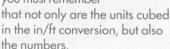

In this calculation you must remember that not only are the units cubed in the in/ft conversion, but also the numbers.

▶ We can then add the cubed in/ft conversion to our density calculation:

$$\text{density} = \frac{\left(\text{mass}\right)}{\left(\text{volume}\right)} = \frac{\left(97.5 \text{ lbs}\right)}{\left(2.00 \text{ ft}^3\right)}\left(\frac{453.59 \text{ g}}{1 \text{ lb}}\right)\left(\frac{1 \text{ ft}}{12 \text{ in}}\right)^3$$

▶ Now we have gotten to the units of g/in^3. Since we do not have the units we are seeking (g/cm^3), we need at least one more step. We can finish the problem in one more step if we know the relationship between inches and centimeters. If we do not know this conversion, we will need more than one additional step. The inch–centimeter relationship is 1 in = 2.54 cm. Since the inches are cubed, we will need to cube this relationship. Our density problem becomes:

$$\text{density} = \frac{\left(\text{mass}\right)}{\left(\text{volume}\right)} = \frac{\left(97.5 \text{ lb}\right)}{\left(2.00 \text{ ft}^3\right)}\left(\frac{453.59 \text{ g}}{1 \text{ lb}}\right)\left(\frac{1 \text{ ft}}{12 \text{ in}}\right)^3\left(\frac{1 \text{ in}}{2.54 \text{ cm}}\right)^3$$

▶ This gives us the desired units of g/cm^3. Since we now have the correct units, the problem is now complete except for calculating the final answer. (Notice that we did not concern ourselves with any actual calculations until now.)

▶ We can now enter the preceding values into a calculator and get:

$$\text{density} = 0.780896 = 0.781 \text{ g/cm}^3$$

▶ If you got 725 instead of 0.781, you forgot to cube both the 12 and 2.54 when performing those steps. Other answers will result if you forget to cube one but not the other.

Careful application of the unit conversion method will make it much easier and quicker to master your chemistry problems.

Studying chemistry requires the student to be an active participant. You must take good notes, study, and, above all, work problems. Regular study every day is far more effective than cramming right before an exam.

EXERCISES

EXERCISE 1-1

You will probably need a calculator to do many of the conversions. This is a good point to make sure you know how to use your calculator efficiently. It is important that you know how to enter numbers in exponential (scientific) notation into your calculator.

1. A substance that has a definite volume and shape is a _____.

2. The abbreviation μ stands for the prefix _____ and has the value _____.

3. The answer to the following calculation should have _____ significant figures.

$$\frac{(1.987)(56.3)}{(4.5)}$$

4. When solving a problem by the unit conversion method, the question mark "?" in the following should be replaced by _____.

$$(1.000 \text{ ft})\left(\frac{-}{?}\right)$$

5. True or false: A substance with variable size and variable shape is a liquid.

6. True or false: The abbreviation *M* refers to *milli-* (0.001).

7. True or false: The result of the following calculation will have three significant figures.

$$3.3 + 4.2 + 4.5 =$$

8. True or false: According to the unit conversion method, the answer to the following calculation is 1728 in^3.

$$\left(1.000 \text{ ft}^3\right)\left(\frac{12 \text{ in}}{1 \text{ ft}}\right)^3 =$$

9. If 2.54 cm = 1 in exactly, which of the following will correctly convert 5.0 in^2 to cm^2?

   a. $\left(5.0 \text{ in}^2\right)\left(\dfrac{2.54 \text{ cm}^2}{1 \text{ in}^2}\right)$

   b. $\left(5.0 \text{ in}^2\right)\left(\dfrac{2.54 \text{ cm}}{1 \text{ in}}\right)^2$

   c. $\left(5.0 \text{ in}\right)^2\left(\dfrac{2.54 \text{ cm}}{1 \text{ in}}\right)^2$

   d. $\left(5.0 \text{ in}^2\right)\left(\dfrac{2.54 \text{ cm}}{1 \text{ in}^2}\right)$

10. If 1 kg is 1000 g, and 1 cm is 0.01 m, which of the following will give the correct result when converting?

   a. $\left(\dfrac{2.75 \text{ g}}{\text{cm}^3}\right)\left(\dfrac{1000 \text{ kg}}{1 \text{ g}}\right)\left(\dfrac{1 \text{ cm}}{0.01 \text{ m}}\right)^3$

   b. $\left(\dfrac{2.75 \text{ g}}{\text{cm}^3}\right)\left(\dfrac{1000 \text{ kg}}{1 \text{ g}}\right)\left(\dfrac{1 \text{ cm}^3}{0.01 \text{ m}^3}\right)$

   c. $\left(\dfrac{2.75 \text{ g}}{\text{cm}}\right)^3\left(\dfrac{1 \text{ kg}}{1000 \text{ g}}\right)\left(\dfrac{1 \text{ cm}^3}{0.01 \text{ m}^3}\right)$

   d. $\left(\dfrac{2.75 \text{ g}}{\text{cm}^3}\right)\left(\dfrac{1 \text{ kg}}{1000 \text{ g}}\right)\left(\dfrac{1 \text{ cm}}{0.01 \text{ m}}\right)^3$

11. Using only the relationships given in this chapter, convert 15.2 in to centimeters.

12. Using only the relationships given in this chapter, convert 15.2 cm to inches.

13. Convert 2.5 ft^3 to in^3.

14. If 1 lb equals 453.59 g, how many grams are in 2.53 lb?

15. How many quarts of liquid are in a 2.0 L soft drink bottle? (1.057 qt = 1 L)

16. How many milliliters of milk are in 1.00 qt of milk?

17. Using the conversion from question 14, calculate the number of kilograms in 2.205 lb.

18. How many square centimeters are in 4.00 square yards? (30.48 cm = 1 ft)

19. How many kilograms are in 3.27 tons? (Use 1 lb = 453.59 g)

20. Using the relationship 2.54 cm = 1 in, calculate the number of square feet in 2.0 m^2.

21. Determine the density, in g/mL, of a rock that weighs 52 g and has a volume of 15 mL.

22. What is the mass of a piece of wood with a volume of 15 cm^3 and a density of 0.8255 g/cm^3?

23. Determine the volume of a 17.5 g sample of a liquid with a density of 0.7826 g/mL.

24. Determine the volume in cubic meters of a 14.2 kg sample of steel with a density of 7.8 g/ cm^3.

25. A 2.00 ft³ sample of water weighs 125 lb. Determine the density of water in grams per cubic centimeter. Use the conversions 1 lb = 453.59 g and 2.54 cm = 1 in.

2 Atoms, Ions, and Molecules

MUST KNOW

⚡ The number of protons determines the identity of an element.

⚡ The naming of chemical compounds follows specific conventions. Knowing how to name chemicals will allow you to work with them effectively and will form part of the foundation of your success in chemistry.

n this chapter we are going to examine the parts of the atom (the subatomic particles—protons, neutrons, and electrons—and where they are found within the atom), what determines the identity of an element (the number of protons), and the systematic arrangement of the elements on the periodic table. Then we will examine the writing and naming of chemical formulas. Let's dive in!

The Subatomic Particles

Our modern model describes the atom as an electrically neutral sphere with a tiny nucleus in the center containing positively charged protons and neutral neutrons. The negatively charged electrons are moving in complex paths outside the nucleus in energy levels at different distances from the nucleus. These subatomic particles have very little mass expressed in grams; so, we often use the unit of an **atomic mass unit (amu** or simply **u)**. An amu is 1/12 the mass of a carbon-12 atom that contains six protons and six neutrons. The following table summarizes the properties of the three subatomic particles.

 IRL Many books and teachers omit the charges on the particle symbols.

Properties of the Subatomic Particles

Particle	Symbol	Charge	Mass (g)	Mass (amu)	Location
proton	p^+	1+	1.673×10^{-24}	1.007	nucleus
neutron	n^0	0	1.675×10^{-24}	1.009	nucleus
electron	e^-	1−	9.109×10^{-28}	5.486×10^{-4}	outside nucleus

Since the atom is electrically neutral, but is composed of charged particles, the number of positively charged protons and negatively charged electrons are equal.

Elements

Now that we know about atoms, their subatomic particles and where those particles are located, it's time to learn about elements. The atoms in an **element** are basically the same, with one possible exception, the number of neutrons. It is the number of protons in an atom that really defines the type of element. For example, in the element sodium, all sodium atoms (Na) contain 11 protons and 11 electrons (to keep the atom neutral), but the number of neutrons may vary. Atoms of the same element (same number of protons) but differing numbers of neutrons are called **isotopes**. The following symbolization is a way to represent a specific isotope:

$$^{A}_{Z}X$$

X represents the element symbol (from the periodic table), Z is the **atomic number**, the number of protons, and A is the **mass number**, the sum of the protons plus neutrons. Subtracting the atomic number from the mass number ($A - Z$) gives the number of neutrons present in this isotope. For example, $^{23}_{11}$Na would contain 11 protons, 11 electrons, and 12 neutrons ($23 - 11$).

BTW
Both the atomic and mass numbers in this representation are counted values and therefore integers.

The Periodic Table

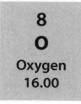

8
O
Oxygen
16.00

The periodic table, such as the one found on page 446 of this book, provides us with a wealth of information. First, it shows the element symbol in the center of the block. Above the element is the element's atomic number, the number of protons in the element's nucleus. Below the element symbol is the element's atomic mass. This mass is not a whole number because it is the average of the masses of all the naturally occurring isotopes considering their abundances (weighted average).

There are several ways that we can use the periodic table to classify the elements. One way is to divide all the elements into three groups: metals, metalloids, and nonmetals. Look at the periodic table in this book. Notice the stair-stepping starting at boron (abbreviated as B) and moving down and to the right. The elements in this area (B, Si, Ge, As, Sb, Te, At) are the **metalloids**. These metalloids have properties of both of the other two groups (metals and nonmetals). Some have unusual electrical properties that make them valuable in the computer and semiconductor industry.

> **BTW**
> The mass number can never be smaller than the atomic number. Never confuse the atomic mass with the mass number.

We classify the elements to the left of this group, excluding the metalloids and hydrogen, as the **metals**. The metals have physical properties that we normally associate with metals in the everyday world—at room temperature they are solids (except for mercury), they have a metallic luster, and they are good conductors of both electricity and heat. They are malleable (capable of being hammered into thin sheets) and ductile (capable of being drawn into thin wires). And as we will see later in this book, the metals tend to lose electrons in chemical reactions.

The other elements, the ones to the right of the metalloids and including hydrogen, are classified as the **nonmetals**. The nonmetals have properties that are opposite of the metals. Many are not solid; they have a dull luster, are nonconductors, and are neither malleable nor ductile. The nonmetals tend to gain electrons in chemical reactions.

> **BTW**
> Even though hydrogen is located on the left side of the periodic table, it is a nonmetal.

Another way of classifying the elements on the periodic table is by the period and group to which they belong. **Periods** are the horizontal rows on the periodic table. They are numbered from 1 to 7. Elements in the same period have consecutive atomic numbers but differ predictably in their chemical properties.

Groups or **families** are the vertical columns on the periodic table. They may be labeled in two ways. The

> **BTW**
> Remember the positions of metals and non-metals relative to the metalloids:
>
> metals ← metalloids → nonmetals

older way involves a Roman numeral and a letter, either A or B. We call the groups labeled with an A the **main-group elements**, while the B groups are the **transition elements**. Two horizontal groups, the **inner transition elements**, belonging to periods 6 and 7 are normally pulled out of the main body of the periodic table and are placed at the bottom of the table.

A newer way of labeling the groups is by consecutive number from left to right, 1 to 18. The older method is still very commonly used and is the labeling method we prefer and will use in this book. However, in some cases, Arabic numerals are substituted for Roman numerals.

Chemists give four main-group families special names:

Group	Family
1A (1)	alkali metals
2A (2)	alkaline earth metals
7A (17)	halogens
8A (18)	noble gases

IRL The noble gases used to be the inert gases because no compounds of these gases were known. However, in 1962, Neil Bartlett created the first "inert gas" compound of xenon, platinum, and fluorine. Shortly afterward scientists started referring to these gases as the noble gases.

Chemical Formulas: Ions and Molecules

Atoms may combine to form compounds. A **compound** is a combination of two or more different elements in a specific ratio. We use a **formula** to represent this compound. H_2O is a formula representing water, a compound composed of two atoms of hydrogen and one atom of oxygen. An **empirical formula** shows the atoms found in the compound and the lowest whole number ratio of those atoms. The empirical formula of water would be

H_2O. However, suppose another compound of hydrogen and oxygen had an empirical formula of HO. This empirical formula tells us that there is only hydrogen and oxygen in the compound and the two atoms are in a 1:1 ratio. The **molecular** or **true formula** shows what atoms we find in the compound and the actual number of each atom. A molecular formula for the empirical formula HO might be H_2O_2, the compound hydrogen peroxide. The **structural formula** shows both the kind and actual number of atoms in the compound but also shows the bonding pattern. We will show you more about structural formulas when you study molecular geometry.

A compound may be formed in two major ways. If a metal is reacting with a nonmetal, then the metal loses one or more electrons while the nonmetal gains those electrons. Ions are formed. An **ion** is an atom or group of atoms that has a charge due to the loss or gain of electrons. If the atom loses electrons, it is left with a positive charge due to having more protons (positive charges) than electrons (negative charges). Ions with a positive charge are **cations**. If the atom gains electrons, it now has more electrons (negative charges) than protons (positive charges) and it is left with a negative charge. Ions with negative charges are **anions**.

Atoms of the main group or representative elements tend to lose or gain enough electrons to achieve the same number of electrons as the noble gas (elements in the column to the far right on the periodic table) closest to the element's atomic number. The alkali metals all tend to lose one electron. For example, sodium and potassium metals would lose a single electron each to form Na^+ and K^+, which have the same number of electrons as the noble gases neon and argon, respectively. The alkaline earth metals lose two electrons. Magnesium and calcium would each lose two electrons to form Mg^{2+} and Ca^{2+}. (Ca and Ca^{2+} are *not* the same; these formulas refer to the element and ion, respectively.) The halogens all tend to gain one electron. Therefore, fluorine and chlorine would form F^- and Cl^-. Oxygen would form O^{2-}.

Sometimes groups of atoms may possess a charge and behave as ions. We call these chemical species **polyatomic ions**. Examples include the ammonium ion, NH_4^+, the nitrate ion, NO_3^-, and the bicarbonate ion, HCO_3^-.

Since opposite charges attract each other, the cations attract the anions, forming an **ionic compound**. Ionic compounds are neutral because the number of positive charges equals the number of negative charges. The potassium cation would attract the chloride anion to form the ionic compound potassium chloride, KCl. We call ionic compounds such as this, salts.

If a nonmetal reacts with another nonmetal, no electrons are lost or gained but are shared, which means there are no ions. We call such compounds **covalent (molecular) compounds**. These compounds contain small units we call **molecules**. Ammonia, NH_3, water, H_2O, and methane, CH_4, are examples of covalent compounds. We will see other examples later.

In the next section, we will examine how to name both salts and molecules.

Metals tend to lose electrons to form cations, while nonmetals tend to gain electrons to form anions.

In an ionic compound, there must be the same number of positive and negative charges. All compounds are neutral.

Naming Compounds

Nomenclature is the term referring to the naming of compounds. In this section, we will investigate how to name inorganic compounds (most compounds not containing carbon). We will see how to name organic compounds (most compounds containing carbon) in Chapter 21. To name compounds correctly, you will need to memorize certain elements and ions, and you will need to apply a few rules.

IRL   Using correct nomenclature helps convince people that you know what you are talking about. In a TV show I recently saw, the commentator referred to the use of silver iodide to seed clouds to create rain. However, he said "silver iodide, AgI_2," which is certainly incorrect nomenclature. He immediately lost a lot of credibility with me and, I imagine, a lot of other science-type people!

We will use a limited set of elements and ions in our examples. You will need to learn additional names. Our limited set begins with the following elements:

Metals		Nonmetals	
sodium	Na	hydrogen	H
magnesium	Mg	nitrogen	N
aluminum	Al	chlorine	Cl
copper	Cu	carbon	C
potassium	K	oxygen	O
calcium	Ca	fluorine	F
iron	Fe	sulfur	S
manganese	Mn		

You will need to commit these names and symbols to memory before attempting to apply the rules of nomenclature. It will also help you locate each of these elements on the periodic table.

You will also need to learn the names, formulas, and charges of the following polyatomic ions:

BTW

Many nonmetals, excluding the noble gases, occur in the form of molecules and not as individual atoms. Examples include H_2, N_2, O_2, F_2, Cl_2, Br_2, and I_2.

nitrate ion	NO_3^-
nitrite ion	NO_2^-
phosphate ion	PO_4^{3-}
sulfate ion	SO_4^{2-}
carbonate ion	CO_3^{2-}
ammonium ion	NH_4^+
bicarbonate ion	HCO_3^-
acetate ion	$C_2H_3O_2^-$

Finally, you will need to memorize the common names of the following molecular compounds:

water	H_2O
ammonia	NH_3
methane	CH_4

When naming compounds containing species not on these lists, it may help to find a chemical species on the list from the same family or a polyatomic ion that is similar.

Some compounds are simple molecules with special names. The short list containing water contains examples of this type of nomenclature. You must simply learn these names; more rules do not alter the fact that H_2O is water.

Some compounds, namely molecular compounds, contain only nonmetals. Normally the compounds you need to name are binary compounds (containing only two elements). If you have highlighted the metalloids on your periodic table, everything to the right of the metalloids is a nonmetal. The following rules apply to both nonmetals and metalloids. The only nonmetal excluded from these nomenclature rules is hydrogen.

When naming a molecular compound, we name each element. The names appear in the same order as they do in the molecular formula. The chemical symbols in the formula are in the order the elements appear on the periodic table. Thus, the element toward the right of the periodic table (excluding the noble gases) will appear toward the right of the formula. If the elements are in the same column, the one nearer the top will be last in the formula.

Hydrogen is nearly always an exception to the rules.

Once you have the symbols in the correct order, you simply write the name of the first element followed by the name of second element. However, you will need to remember that the name of the second element will be changed to end with an *-ide* suffix. Thus:

oxygen → oxide

nitrogen → nitride

fluorine → fluoride

chlorine → chloride

sulfur → sulfide

carbon → carbide

To complete the name, it is necessary to add prefixes to indicate the number of atoms. We will use the following list:

1 = mono

2 = di

3 = tri

4 = tetra

5 = penta

6 = hexa

7 = hepta

8 = octa

9 = nona

10 = deca

The final -o or -a may be dropped if the name of the element begins with a vowel. The prefix *mono-* is being used less often, so you may only see it in a very small number of compounds such as carbon monoxide, CO.

EXAMPLE

▶ Let's use the following compounds as examples: CS_2, ClF_3, and N_2O_5. You should locate the elements in each compound on the periodic table to confirm the order they appear in the formula. The next step is to write the name of each element in the same order that they appear in the formula:

CS_2 ➡ carbon sulfur

ClF_3 ➡ chlorine fluorine

N_2O_5 ➡ nitrogen oxygen

▶ In each case, we need to change the name of the second element to one with an *-ide* suffix:

CS_2 ➡ carbon sulfide

ClF_3 ➡ chlorine fluoride

N_2O_5 ➡ nitrogen oxide

▶ Next, we need to add a prefix in those cases in which there is more than one of a particular atom:

CS_2 ➡ carbon disulfide

ClF_3 ➡ chlorine trifluoride

N_2O_5 ➡ dinitrogen pentoxide

BTW

Notice that in these cases, the mono- *prefix was not used.*

▶ Reversing this procedure will allow us to write a formula from a name. Let's try this with carbon tetrachloride and oxygen difluoride. The formulas for these two compounds are CCl_4 and OF_2, respectively.

The presence of a metal in a compound indicates that ions are probably present. (Recall that, other than hydrogen, all elements to the left of the metalloids on the periodic table are metals.) Metals may combine with nonmetals or with polyatomic ions. We will begin with binary compounds containing metals. The naming of a metal/nonmetal compound is like naming molecular compounds except that you do not use prefixes. (There are a few archaic names, which do include prefixes in their name.) In these compounds, the cation will always appear first in both the name and the formula.

Without using prefixes, we need another method of determining the number of each type of atom present. This method depends on the charges of each ion. We can predict the charge of the ion from the element's position on the periodic table. Unfortunately, this method does not work very well for the transition elements, so we will save a discussion of transition metal compounds until later. For the main group or representative elements, we simply find the charges of the cations by counting from left to right on the periodic table, skipping the transition metals.

If we locate the metals from our list on the periodic table, we see:

Na   Mg   Al
K   Ca

Sodium and potassium are in the first column, so they, and any other metal in this column, should form a cation with a +1 charge. Magnesium and calcium are in the second column, thus they, and any other metal in the same column, should form a cation with a +2 charge. Aluminum is in the third column (skipping the transition metals), so that aluminum, and any other metal in the same column, should form a cation with a +3 charge. Thus, the cations of these five metals are: Na^+, K^+, Mg^{2+}, Ca^{2+}, and Al^{3+}.

Nonmetals are the opposite of metals. Thus, while metals give cations (positive ions), nonmetals will yield anions (negative ions), which are the opposite of cations. We can predict the charge on an anion from the position of the nonmetal on the periodic table. Since nonmetals are the opposite of metals, we will need to count in the opposite direction. The noble gases are zero. The next column to the left will be −1, then −2, and then −3. The nonmetals on our list should form the following anions:

N^{3-} O^{2-} F^- Ne^0
           S^{2-} Cl^- Ar^0

Ions always have charges; compounds never have charges. How can we combine charged ions and end up with no charge? The answer: add equal numbers of positive and negative charges. The sum of equal numbers of positive charges and negative charges is zero. If the total of the charges is not zero, you do not have a compound.

If the metal and the nonmetal have the same magnitude of charge, you only need one of each to produce a compound. Examples are:

$$Na^+ + F^- \rightarrow NaF$$
$$Ca^{2+} + S^{2-} \rightarrow CaS$$
$$Al^{3+} + N^{3-} \rightarrow AlN$$

When the magnitudes of the charges differ, the numbers of atoms will no longer be the same. For example, if we combine calcium and chlorine, calcium produces the Ca^{2+} cation and chlorine produces the Cl^- anion. To balance the +2 charge of the calcium, we need a −2 charge from the chloride. Since each chloride only supplies a −1, we need two chloride ions to get the needed −2 charge. The formula needs one Ca^{2+} and 2 Cl^- giving $CaCl_2$. Now try this with aluminum and oxygen and see if you can get the correct formula (Al_2O_3).

The names of the metal/nonmetals in this section are:

NaF = sodium fluoride

AlN = aluminum nitride

Al_2O_3 = aluminum oxide

CaS = calcium sulfide

$CaCl_2$ = calcium chloride

The presence of a polyatomic ion in a compound requires substituting the name of the polyatomic ion for the name of the cation or anion in a comparable metal/nonmetal compound. We will still not use prefixes.

> ▶ Let's try an example. What is the formula of potassium phosphate?
>
> ▶ We can use our method for assigning the charge of cations to predict that potassium will be present as a K^+ ion. You should already know that the phosphate ion is PO_4^{3-} (from the list earlier in this chapter). The difference in charges shows us that it will take three potassium ions to balance the charge on the phosphate ion. Therefore, the formula of potassium phosphate is K_3PO_4.

The transition metals, such as manganese, iron, and copper on our list, require additional consideration. Most transition metals, and the elements located around lead, Pb, on the periodic table, can form cations of different charges. For this reason, we need additional information from the name of the compound containing these metals.

We will use the three transition metals from our list to illustrate how we need to treat these metals in compounds. Manganese usually forms Mn^{2+} and Mn^{3+} ions. Iron commonly forms either Fe^{2+} or Fe^{3+} ions, and copper commonly forms Cu^+ and Cu^{2+}. (Later you will see how to predict why these metals form these ions.) In older literature, it was necessary to memorize a separate name for each ion. Using this method, Fe^{2+} is the ferrous ion, Fe^{3+} is the ferric ion, Cu^+ is the cuprous ion, Cu^{2+} is the cupric ion, Mn^{2+} is the manganous ion, and Mn^{3+} is the manganic ion. In all cases, the ion with the greater charge has the *-ic* suffix.

To avoid the necessity of memorizing a separate name for each ion, we can use the Stock system. In the Stock system, the charge of the cation appears as a Roman numeral immediately after the name of the element. Using the Stock system, we write Fe^{2+} as the iron(II) ion, and Cu^+ as the copper(I) ion. Other than the necessity of indicating the charges, there are no differences between the naming of transition metal compounds and other compounds of the metals. So, while KCl is potassium chloride, CuCl is copper(I) chloride.

Finally, we need to consider compounds containing the nonmetal hydrogen. Remember that hydrogen is an exception. In simple binary compounds with nonmetals, we treat hydrogen as a metal. As a "metal" in the first column, it should have a +1 charge. Thus, H_2S is hydrogen sulfide.

Many hydrogen compounds are acids (defined later). Acids require a different system of nomenclature than other compounds. There are two types of acids: binary acids and ternary acids. All acids include the word *acid* in their name. Binary acids contain only two elements: one of which is hydrogen and the other a nonmetal. When naming binary acids, the prefix *hydro-* appears before the root name of the nonmetal, and there will be an *-ic* suffix. As an acid, H_2S would be hydrosulfuric acid. We used H_2S to illustrate the fact that there can be two possible names, hydrogen sulfide or hydrosulfuric acid. The "correct" name depends on the circumstances. Technically, the name hydrosulfuric acid only applies to aqueous solutions of hydrogen sulfide, $H_2S(aq)$.

The ternary acids consist of hydrogen combined with a polyatomic ion. As with binary acids, the word *acid* must appear in the name. The remainder of the name of the acid will come from the name of the polyatomic ion. Ions ending in *-ite* have these three letters changed to *-ous*, and ions ending in *-ate* have these three letters changed to *-ic*. Thus, the nitrite ion, NO_2^-, becomes nitrous acid, HNO_2, and the nitrate ion, NO_3^-, becomes nitric acid, HNO_3. Some ions require a little more work; for example, the sulfate ion, SO_4^{2-}, becomes sulfuric acid, H_2SO_4.

EXERCISES

EXERCISE 2-1

Answer the following questions.

1. The heaviest subatomic particle is the _____.

2. The only subatomic particle found outside the nucleus is the _____.

3. Elements toward the left side of the periodic table are _____.

4. The horizontal rows on the periodic table are the _____.

5. Ions with a positive charge are _____.

6. True or False: A neutron is heavier than a proton.

7. True or False: The mass number of an element appears on the periodic table.

8. True or False: The metalloids look like metals, but otherwise they have no similarity to metals.

9. True or False: A group of atoms with an overall positive or negative charge is an example of a compound.

10. Elements in the first column on the periodic table (1A or 1) usually have what charge in a compound?
    a. +1
    b. −1
    c. +2
    d. −2
    e. +3
    f. 0

11. Elements in the second column on the periodic table (2A or 2) usually have what charge in a compound?
    a. +1
    b. −1
    c. +2
    d. −2
    e. +3
    f. 0

12. Elements in the second to last column on the periodic table (7A or 17) usually have what charge in a compound?
    a. +1
    b. −1
    c. +2
    d. −2
    e. +3
    f. 0

13. Elements in the third to last column on the periodic table (6A or 16) usually have what charge in a compound?
    a. +1
    b. −1
    c. +2
    d. −2
    e. +3
    f. 0

14. The sodium ions in sodium chloride, NaCl, have what charge?
    a. +1
    b. −1
    c. +2
    d. −2
    e. +3
    f. 0

15. The oxide ions in calcium oxide, CaO, have what charge?
    a. +1
    b. −1
    c. +2
    d. −2
    e. +3
    f. 0

16. The aluminum atoms in a piece of aluminum foil have what charge?
    a. +1
    b. −1
    c. +2
    d. −2
    e. +3
    f. 0

17. The iron ions in iron(III) phosphide, FeP, have what charge?
    a. +1
    b. −1
    c. +2
    d. −2
    e. +3
    f. 0

18. The simplest formula for a compound containing Mn^{4+} and O^{2-} is:
    a. MnO
    b. Mn_2O_4
    c. Mn_4O_2
    d. MnO_2
    e. none of these

19. The correct formula for dichlorine trioxide is:
    a. Cl_2O
    b. Cl_2O_6
    c. Cl_2O_3
    d. $Cl_2O_3^{2-}$
    e. none of these

20. Complete the following table.

	$^{23}_{11}Na$	$^{40}_{19}K^+$	$^{53}_{25}Mn^{2+}$		
protons	11			54	
neutrons					42
electrons		18			
charge	0	+1		0	−3
mass number	23			130	75

21. Name each of the following compounds.
    a. Na_2S
    b. $MgCl_2$
    c. AlF_3
    d. CuO
    e. KH

22. Give the formula of each of the following compounds.
    a. calcium chloride
    b. sodium oxide
    c. aluminum nitride
    d. manganese(II) fluoride
    e. manganese(III) fluoride

23. Name each of the following compounds.
    a. NO_2
    b. CO
    c. H_2S
    d. Cl_2O_5
    e. CH_4

24. Give the formula of each of the following compounds.
    a. hydrogen chloride
    b. sulfur dioxide
    c. sulfur trioxide
    d. dinitrogen pentoxide
    e. ammonia

25. Name each of the following compounds.
    a. $NaNO_3$
    b. K_2CO_3
    c. $CaSO_4$
    d. $(NH_4)_3PO_4$
    e. $Fe(NO_2)_2$

26. Give the formula of each of the following compounds.
    a. sodium phosphate
    b. ammonium bicarbonate
    c. magnesium nitrite
    d. copper(I) chloride
    e. iron(III) oxide

27. Name each of the following acids.
    a. HF
    b. HNO_3
    c. H_3PO_4
    d. H_2CO_3
    e. H_2SO_4

28. Give the formula of each of the following acids.
    a. nitrous acid
    b. hydrosulfuric acid
    c. sulfuric acid
    d. hydrochloric acid
    e. nitric acid

29. Locate each of the following elements on the periodic table and label it as a metal or a nonmetal.
    a. Cs
    b. Ra
    c. Tl
    d. P
    e. Se
    f. I

30. Based on their positions on the periodic table, assign charges to the ions that the elements in question 29 might form.

 # Moles, Stoichiometry, and Equations

 ## MUST KNOW

 The mole is a fundamental unit of measurement of substances. A mole has 6.022×10^{23} particles (Avogadro's number) and is also the atomic or formula mass of a substance expressed in grams.

 We can interpret the coefficients in the balanced chemical equation as a mole relationship as well as a particle one. Using these relationships, we can determine how much reactant is needed and how much product can be formed—the stoichiometry of the reaction.

Mass data allows us to determine the percentage of each element in a compound and the empirical and molecular formulas.

This is going to prove to be a critical chapter in our study of chemistry. To be able to grasp the mole concept, it will be necessary to learn about balancing equations and the mole–mass relationships (stoichiometry) inherent in these balanced equations. We will learn, given amounts of reactants, how to determine which one limits the amount of product formed, as well as how to determine the empirical and molecular formulas of compounds. All of these will depend on the mole concept. It's a lot—but doable!

Balancing Chemical Equations

A chemical equation represents a chemical change that is taking place. On the left side of the reaction arrow are the **reactants**, the chemical substances that are changed. On the right of the reaction arrow are the reaction **products**, the new substances formed. Sometimes additional information appears above or below the reaction arrow.

The **law of conservation of mass** states that the total mass remains unchanged. This means that the total mass of the atoms of each element represented in the reactants must appear as products. To indicate this, we must balance the reaction. When balancing chemical equations, it is important to realize that you cannot change the formulas of the reactants and products; the only things you may change are the coefficients in front of the reactants and products. The coefficients indicate how many of each chemical species react or form. A balanced equation has the same number of each type of atom present on both sides of the equation and the coefficients are present in the lowest whole number ratio. For example, iron metal reacts with oxygen gas to form rust, iron(III) oxide. We may represent this reaction by the following balanced equation:

$$4 \ Fe(s) + 3 \ O_2(g) \rightarrow 2 \ Fe_2O_3(s)$$

IRL The oxidation of iron to form rust is not a good thing. It destroys the integrity of the metal. However, the oxidation of aluminum to form Al_2O_3 is normally considered a good thing. Aluminum oxide is a very hard substance that protects the rest of the aluminum metal from further reacting with the air.

This equation tells us that four iron atoms react with three oxygen molecules to form two rust formulas. Note that the number of iron atoms and oxygen atoms are the same on both sides of the equation ($2 \times Fe_2O_3 =$ 4 Fe and 6 O). Let's see how we went about arriving at the balanced equation for the rusting of iron.

EXAMPLE

▶ First, write an unbalanced equation, showing the reactants and products:

$$\underline{\quad} \ Fe(s) + \underline{\quad} \ O_2(g) \rightarrow \underline{\quad} \ Fe_2O_3(s)$$

▶ The blanks indicate where we will add coefficients. When balancing, you can make no changes other than placing numbers in these blanks. Note that there is one iron on the reactant side and two on the product side. To balance the iron, we need a coefficient of 2 in front of the Fe on the reactant (left) side:

$$2 \ Fe(s) + \underline{\quad} \ O_2(g) \rightarrow \underline{\quad} \ Fe_2O_3(s)$$

▶ The iron atoms are balanced, but there are two oxygen atoms on the reactant side and three on the product side. The oxygens enter the reaction in groups of two, O_2. We will balance the oxygen atoms by placing a coefficient of 3/2 in front of the O_2 on the left:

BTW
You can't change the formulas of the reactants or products, just the coefficients.

$$2 \ Fe(s) + 3/2 \ O_2(g) \rightarrow \underline{\quad} \ Fe_2O_3(s)$$

▶ That will give us three oxygens on both sides ($3/2 \times 2 = 3$), but we must have coefficients that are whole numbers. The easiest way to achieve this is to multiply everything by 2:

$$2 \times [2\ Fe(s) + 3/2\ O_2(g) \rightarrow 1\ Fe_2O_3(s)],$$

giving the balanced equation:

$$4\ Fe(s) + 3\ O_2(g) \rightarrow 2\ Fe_2O_3(s)$$

BTW

If a coefficient is 1, we commonly omit it.

We can balance most simple reactions by this trial and error method, by inspection, but some reactions, redox reactions, often require a different system of rules. We will show you how to balance these redox reactions in Chapter 18.

BTW

Make sure that all atoms balance and the coefficients are in the lowest whole number ratio.

Avogadro's Number and Molar Mass

The **mole** (**mol**) is the amount of a substance that contains the same number of particles as atoms in exactly 12 grams of carbon-12. This number of particles (atoms, molecules, or ions) per mole is **Avogadro's number** and is numerically equal to 6.022×10^{23} particles. The mole is simply a term that represents a certain number of particles, like a dozen or a pair. The mole also represents a certain mass of a chemical substance.

The substance's **molar mass** is the mass in grams of the substance that contains one mole of that substance. In the previous chapter, we described the atomic mass of an element in terms of atomic mass units (amu). This was the mass associated with an individual atom. At the microscopic level, we can calculate the mass of a compound by simply adding together the masses in atomic mass units of the individual elements in the compound. However, at the macroscopic level, we use the unit of grams to represent the quantity of a mole. (The atomic masses are listed on the periodic table.)

BTW

6.022×10^{23} particles = 1 mole = molar mass in grams

This relationship in the previous BTW sidebar gives a way of converting from grams to moles to particles and vice versa. If you have any one of the three quantities, you can calculate the other two. For example, the molar mass of iron(III) oxide, Fe_2O_3 (rust), is 159.689 g/mol = [(2 × 55.846 g/mol for Fe) + (3 × 15.999 g/mol for O)]. Therefore, if we had 50.00 g of iron(III) oxide, we could calculate both the number of moles and the number of particles present.

$$\left(50.00 \text{ g Fe}_2O_3\right)\left(\frac{1 \text{ mol Fe}_2O_3}{159.689 \text{ g}}\right) = 0.313108605 \text{ mol}$$

$$= 0.3131 \text{ mol Fe}_2O_3$$

$$\left(0.313108605 \text{ mol Fe}_2O_3\right)\left(\frac{6.022 \times 10^{23} \text{ particles}}{1 \text{ mol Fe}_2O_3}\right)$$

$$= 1.886 \times 10^{23} \text{ Fe}_2O_3 \text{ particles}$$

Notice that if we had grams and wanted just particles, we still would need to incorporate the mole relationship.

BTW

If you are starting with either grams or particles, you will need to calculate moles.

Moles and Stoichiometry

As we mentioned previously, the balanced chemical equation not only indicates what chemical species are the reactants and what the products are, but it also indicates the relative ratio of reactants and products. Consider the balanced equation for the rusting of iron:

$$4 \text{ Fe}(s) + 3 \text{ O}_2(g) \rightarrow 2 \text{ Fe}_2O_3(s)$$

This balanced equation can be read as: 4 iron atoms react with 3 oxygen molecules to produce 2 iron(III) oxide units. However, the coefficients can stand not only for the number of atoms or molecules (microscopic level) but they can also stand for the number of *moles* of reactants or products (at the macroscopic level). So, the equation can also be read as: 4 moles of iron react with 3 moles of oxygen to produce 2 moles of iron(III) oxide. In addition, if we know the number of moles, the number of grams or molecules may be calculated. This is **stoichiometry**, the calculation of the amount (mass, moles, particles) of one substance in the chemical equation from another. The coefficients in the balanced chemical equation define the mathematical relationship between the reactants and products and allow the conversion from moles of one chemical species in the reaction to another.

EXAMPLE

▶ Consider the preceding rusting process. How many grams of rust (Fe_2O_3) could form in the reaction of 20.0 mol of iron with excess oxygen?

▶ You are starting with moles of iron and want grams of Fe_2O_3, so we'll first convert from moles of iron to moles of Fe_2O_3 using the ratio of moles of Fe_2O_3 to moles of iron as defined by the balanced chemical equation, and then use the molar mass of Fe_2O_3 as determined above:

> **BTW**
>
> Before any reaction calculations can be done, you **must** have a balanced chemical equation!

$$(20.0 \text{ mol Fe})\left(\frac{2 \text{ mol Fe}_2\text{O}_3}{4 \text{ mol Fe}}\right)\left(\frac{159.689 \text{ g Fe}_2\text{O}_3}{1 \text{ mol Fe}_2\text{O}_3}\right)$$

$$= 1.60 \times 10^3 \text{ g Fe}_2\text{O}_3$$

▶ The ratio of 2 mol Fe_2O_3 to 4 mol Fe is called a stoichiometric ratio, which comes directly from the coefficients in the balanced chemical equation.

▶ Suppose you also wanted to know how many grams of oxygen it would take to react with 19.9 mol of iron. All you would need to do would be to change the stoichiometric ratio and the molar mass:

> **BTW**
>
> In working stoichiometry problems, you will need the balanced chemical equation. In addition, if the problem involves a quantity other than moles, you will need to convert to moles.

$$(19.9 \text{ mol Fe})\left(\frac{3 \text{ mol O}_2}{4 \text{ mol Fe}}\right)\left(\frac{31.998 \text{ g O}_2}{1 \text{ mol O}_2}\right) = 478 \text{ g O}_2$$

▶ Notice that this new stoichiometric ratio also came from the balanced chemical equation.

Limiting Reactant and Percent Yield

In the preceding examples, we indicated that one reactant was present in excess. The other reactant is consumed and there would be some of the reactant in excess left over. The first reactant to react completely is the **limiting reactant** (also called *limiting reagent*). This reactant really determines the amount of product formed. There are, in general, two ways to determine which reactant is the limiting reactant:

- ■ We assume each reactant, in turn, to be the limiting reactant and we calculate the amount of product that forms. The reactant that yields the *smallest* amount of product is the limiting reactant.

- ■ We calculate the mole-to-coefficient ratio of each reactant by dividing the moles of that reactant by its coefficient in the balanced chemical equation. The reactant that has the smallest mole-to-coefficient ratio is the limiting reactant. Many of us use this method.

EXAMPLE

▶ Let us consider the rusting reaction once more. Suppose that 50.0 g of iron and 40.0 g of oxygen react. Calculate the number of grams of iron(III) oxide that could be formed.

▶ First, write the balanced chemical equation:

$$4 \text{ Fe}(s) + 3 \text{ O}_2(g) \rightarrow 2 \text{ Fe}_2\text{O}_3(s)$$

▶ Next, convert the grams of each reactant to moles:

$$(50.0 \text{ g Fe})\left(\frac{1 \text{ mol Fe}}{55.846 \text{ g Fe}}\right) = 0.89531927 \text{ mol Fe (unrounded)}$$

$$(40.0 \text{ g O}_2)\left(\frac{1 \text{ mol O}_2}{31.998 \text{ g O}_2}\right) = 1.2500078 \text{ mol O}_2 \text{ (unrounded)}$$

▶ Divide each by the coefficient in the balanced chemical equation. The smallest result is the limiting reactant:

$$\text{Fe} \rightarrow 0.89531927 \text{ mol Fe}/4 = 0.2238298 \text{ mol/coefficient}$$
$$\text{(limiting reactant)}$$

$$\text{O}_2 \rightarrow 1.250078 \text{ mol O}_2/3 = 0.4166927 \text{ mol/coefficient}$$

▶ Finally, base the stoichiometry of the reaction on the limiting reactant:

$$(50.0 \text{ g Fe})\left(\frac{1 \text{ mol Fe}}{55.846 \text{ g Fe}}\right)\left(\frac{2 \text{ mol Fe}_2\text{O}_3}{4 \text{ mol Fe}}\right)\left(\frac{159.689 \text{ g Fe}_2\text{O}_3}{1 \text{ mol Fe}_2\text{O}_3}\right)$$

$$= 71.5 \text{ g Fe}_2\text{O}_3$$

In the previous problem, the amount of product calculated based upon the limiting reactant concept is the maximum amount of product that will form from the specified amounts of reactants. This maximum amount of product is the **theoretical yield**. However, rarely is the amount formed (the **actual yield**) the same as the theoretical yield. Normally it is less. There are many reasons for this, but the principal one is that most reactions do not go to completion; they establish an equilibrium system (see Chapter 14 for a discussion on chemical equilibrium). For whatever reason, not as much product as expected is formed. We can judge the efficiency of the reaction by calculating the percent yield. The **percent yield (% yield)** is the actual

BTW

Anytime you are given the amounts (grams, moles, etc.) of more than one reactant, you will need to determine the limiting reactant.

yield divided by the theoretical yield and the result multiplied by 100% to generate a percentage:

$$\% \text{ yield} = \frac{\text{actual yield}}{\text{theoretical yield}} \times 100\%$$

EXAMPLE

▶ In the previous problem, we calculated that 71.5 g of Fe_2O_3 could be formed. Suppose that after the reaction we found that only 62.3 g of Fe_2O_3 formed. Calculate the percent yield.

$$\% \text{ yield} = \frac{62.3 \text{ g}}{71.5 \text{ g}} \times 100\% = 87.1\%$$

Percent Composition and Empirical Formulas

If we know the formula of a compound, it is a simple task to determine the percent composition of each element present.

EXAMPLE

▶ For example, suppose you wanted the percentage carbon and hydrogen in methane, CH_4. First, calculate the molecular mass of methane:

$$1 \text{ mol } CH_4 = 1 \text{ mol } C + 4 \text{ mol } H$$

▶ Substituting the molar masses (from the periodic table):

$$1 \text{ mol } CH_4 = 1(12.01 \text{ g/mol}) + 4(1.008 \text{ g/mol}) = 16.042 \text{ g/mol}$$

▶ (This is an intermediate calculation—don't worry about significant figures yet.)

$$\% \text{ carbon} = [\text{mass C / mass } CH_4] \times 100\%$$
$$= [12.01 \text{ g/mol / 16.042 g/mol}] \times 100\% = 74.87\% \text{ C}$$

% hydrogen = [mass H / mass CH_4] × 100%
= [4(1.008 g/mol) / 16.042 g/mol] × 100% = 25.13% H

In the preceding problem, we determined the percentage data from the chemical formula. We can determine the empirical formula if we know the percent compositions of the various elements. The **empirical formula** tells us what elements are present in the compound and the simplest whole-number ratio of elements. The data may be in terms of percentage, or mass, or even moles. However, the procedure is still the same: convert each element to moles, divide each by the smallest, and then use an appropriate multiplier if necessary. We can then determine the empirical formula mass. If we know the actual molecular mass, dividing the molecular formula mass by the empirical formula mass gives an integer (rounded if needed) that we can multiply by each of the subscripts in the empirical formula. This gives the **molecular (actual) formula**, which tells what elements are in the compound and the actual number of each.

BTW

As a good check, add all the percentages together. They should equal to 100 percent or be very, very close. If not, you have made an error.

EXAMPLE

An analysis of a gas sample found 2.34 g of nitrogen and 5.34 g of oxygen present. What was the empirical formula of this gas?

$$(2.34 \text{ g N})\left(\frac{1 \text{ mol N}}{14.01 \text{ g}}\right) = 0.167 \text{ mol N} \qquad \left(\frac{0.167}{0.167}\right) = 1.00$$

$$(5.34 \text{ g O})\left(\frac{1 \text{ mol O}}{16.00 \text{ g}}\right) = 0.334 \text{ mol O} \qquad \left(\frac{0.334}{0.167}\right) = 2.00$$

The data indicate that the nitrogen and oxygen are present in a 1:2 ratio. Therefore, the empirical formula is NO_2.

If we determined that the actual molecular mass of this gas was 92.00 g/mol, what is the molecular formula?

▶ The empirical formula mass is 46.00 g/mol. Dividing the actual molar mass by the empirical molar mass gives:

$$\frac{92.00 \text{ g / mol}}{46.00 \text{ g / mol}} = 2$$

▶ Therefore, the molecular formula is twice the empirical formula or N_2O_4.

Confronting Mole Problems

The mole is the most important concept in this chapter. Nearly every problem associated with this material requires moles in at least one of the steps. You should get into the habit of automatically looking for moles. There are several ways of finding the moles of a substance. You may determine the moles of a substance from a balanced chemical equation. You may determine moles from the mass and molar mass of a substance. You may determine moles from the number of particles and Avogadro's number. You may find moles from the moles of another substance and a mole ratio. Later in this book, you will find even more ways to determine moles. In some cases, you will be finished when you find moles; in other cases, finding moles is only one of the steps in a longer problem.

EXAMPLE

▶ Let's use the following chemical equation to examine the different ways of finding moles:

$$H_2(g) + Cl_2(g) \rightarrow 2 \text{ HCl}(g)$$

▶ It is possible to consider a balanced chemical equation, such as this one, on many levels. One of these levels is the mole level. At this level, we can get the moles just by reading the equation in terms of moles. 1 mol of H_2 plus 1 mol of Cl_2 yields 2 mol of HCl.

▶ We can also find the moles of, for example, chlorine, from the mass and the molecular weight. So, if we have 175 g of chlorine, how many moles of chlorine do we have?

$$\text{mol Cl}_2 = \left(175 \text{ g Cl}_2\right)\left(\frac{1 \text{ mol Cl}_2}{70.9 \text{ g Cl}_2}\right) = 2.4683 = 2.47 \text{ mol Cl}_2$$

▶ The value 70.9 is the molar weight (mass) of the diatomic chlorine obtained by multiplying the atomic mass of chlorine (from the periodic table) by 2.

▶ If we want to find the moles of a substance from the number of particles, we need to use Avogadro's number. For example, let's find the moles of hydrogen chloride in 2.62×10^{24} molecules of HCl.

$$\text{mol HCl} = \left(2.62 \times 10^{24} \text{ molecules HCl}\right)\left(\frac{1 \text{ mol HCl}}{6.022 \times 10^{23} \text{ molecules HCl}}\right)$$
$$= 4.3507 = 4.35 \text{ mol HCl}$$

▶ Let's examine the last of the common ways to find moles. This method requires that you have the moles of one substance to get the moles of another substance. The moles of the first substance may be given to you or you may have to use one of the above methods to determine the initial mole value. Suppose we wished to know the number of moles of HCl that we might form from 2.2 mol of H_2.

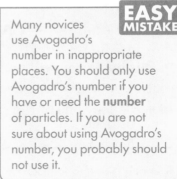

EASY MISTAKE

Many novices use Avogadro's number in inappropriate places. You should only use Avogadro's number if you have or need the **number** of particles. If you are not sure about using Avogadro's number, you probably should not use it.

$$\text{mol HCl} = \left(2.2 \text{ mol H}_2\right)\left(\frac{2 \text{ mol HCl}}{1 \text{ mol H}_2}\right)$$
$$= 4.4 \text{ mol HCl}$$

▶ The term ($2 \text{ mol HCl}/1 \text{ mol H}_2$) is a mole ratio. We got this mole ratio directly from the **balanced chemical equation**. The balanced chemical equation has a 2 in front of the HCl, thus we have the same number in front of the "mol HCl." The balanced chemical equation has an understood 1 in front of the H_2; for this reason the same value belongs in front of the "mol H_2." The values in the mole ratio are exact numbers, and as such, do not affect the significant figures.

EXTRA HELP

Let's use the information from this chapter in one more problem. This will be a long problem that will probably have more facets to it than any other problem you may see, but it will help you see all sides of mole problems. To simplify this problem, consider doing it in a series of small steps instead of trying to do it in one or two large steps.

EXAMPLE

▶ We will begin by stating the problem: solid iodine, I_2, will react with fluorine gas, F_2, to form gaseous iodine pentafluoride, IF_5. In one experiment, a scientist mixed 75.0 g of iodine with 4.00×10^{23} molecules of fluorine and allowed them to react. What is the percent yield if the reaction produced 45.2 g of iodine pentafluoride?

▶ There are many important aspects of this problem. We must begin by extracting the information. We have enough information to begin writing a reaction and several quantities (with associated units listed below the appropriate substance):

$$I_2(s) \quad + \quad F_2(g) \quad \longrightarrow \quad IF_5(g) \quad \text{(unbalanced)}$$

75.0 g	4.00×10^{23}	45.2 g
	molecules	? percent yield

▶ Now that we pulled all the information from the problem, we no longer need the original problem. Having the pertinent information, we can go to the second step that involves the equation.

BTW

In order to work any problem dealing with a chemical reaction, you **must** have a balanced chemical equation.

▶ Since our initial equation is not balanced, we need to balance it.

$$I_2(s) \quad + \quad F_2(g) \quad \longrightarrow \quad IF_5(g)$$

$$\downarrow \qquad\qquad \downarrow \qquad\qquad\qquad \downarrow$$

$$I_2(s) \quad + \quad 5\,F_2(g) \quad \longrightarrow \quad 2\,IF_5(g)$$

75.0 g	4.00×10^{23}	45.2 g
	molecules	? percent yield

▶ The most important concept when working stoichiometry problems such as this one is moles. We must have moles to proceed. The mole determination of iodine will involve the molar mass of iodine (2×126.9 g/mol), while the mole determination of fluorine will involve Avogadro's number (since we have the number of fluorine molecules). We can find the moles of each as follows:

$$\text{mol } I_2 = \left(75.0 \text{ g } I_2\right)\left(\frac{1 \text{ mol } I_2}{253.8 \text{ g } I_2}\right) = 0.295508 \text{ mol } I_2 \text{ (unrounded)}$$

$$\text{mol } F_2 = \left(4.00 \times 10^{23} \text{ molecules } F_2\right)\left(\frac{1 \text{ mol } F_2}{6.022 \times 10^{23} \text{ molecules } F_2}\right)$$
$$= 0.66423 \text{ mol } F_2 \text{ (unrounded)}$$

▶ We could determine the moles of IF_5 at this point; however, we will forego that pleasure until we finish with the reactants.

▶ We now have the moles of the two reactants. This should ring a bell: any time you have the quantity of more than one reactant, you need to know which is the limiting reactant. For this reason, our next step will be to determine which of our reactants limits the reaction. We can determine which reactant is the limiting reactant by dividing each of the moles by the coefficient of each reactant from the balanced chemical equation. (This is one place where, if we did not balance the equation, we would be in trouble.)

$$I_2 = \frac{0.295508 \text{ mol } I_2}{1} = 0.295508 \text{ (unrounded)}$$

$$F_2 = \frac{0.66423 \text{ mol } F_2}{5} = 0.132846 \text{ (unrounded)}$$
$$\text{(limiting reactant)}$$

▶ The quantity (mole-to-coefficient ratio) of F_2 is smaller than that of I_2; therefore, fluorine is the limiting reactant (reagent). Once we know the limiting reactant, *all* remaining calculations will depend on the limiting reactant.

▶ Now we can find the maximum number of moles of IF_5 that can be produced (theoretical yield). We will begin with our limiting reactant rather than using the 45.2 g of IF_5 given (actual yield). Remember, the limiting reactant is our key, and we don't want to lose our key once we have found it.

▶ To find the moles of IF_5 from the limiting reactant, we need to use a mole ratio derived from information in the balanced chemical equation. (This is another place where, if we had not balanced the equation, we would be in trouble.)

$$\text{mol } IF_5 = \left(0.66423 \text{ mol } F_2\right)\left(\frac{2 \text{ mol } IF_5}{5 \text{ mol } F_2}\right) = 0.132846$$

▶ We will not actually calculate the moles of IF_5 yet; we will simply save this setup.

▶ The problem asks for the percent yield. Recall the definition of percent yield:

$$\% \text{ yield} = \frac{\text{actual yield}}{\text{theoretical yield}} \times 100\%$$

▶ In this problem, the actual yield is the amount of product found by the scientist (45.2 g IF_5); therefore, we need the theoretical yield to finish the problem. Since our actual yield has the units "g IF_5," our theoretical yield must have identical units. To determine the grams of IF_5 from the moles of limiting reactant, we need the molar mass of IF_5 [126.9 g I/mol I + 5 (19.00 g F/mol F) = 221.9 g IF_5/mol]:

$$\text{mass } IF_5 = \left(0.66423 \text{ mol } F_2\right)\left(\frac{2 \text{ mol } IF_5}{5 \text{ mol } F_2}\right)\left(\frac{221.9 \text{ g } IF_5}{1 \text{ mol } IF_5}\right)$$

$$= 58.957 \text{ g } IF_5 \text{ (unrounded)}$$

▶ We can enter this theoretical yield (58.957 g IF_5), along with the actual yield (45.2 g IF_5), into the percent yield definition:

$$\% \text{ yield} = \frac{45.2 \text{ g } IF_5}{58.957 \text{ g } IF_5} \times 100\% = 76.666 = 76.7\%$$

Well, you made it through this difficult problem. Pat yourself on the back and forge ahead, but don't hesitate to come back and review this problem as needed.

BTW

All the units in the percent yield calculation, except %, must cancel. This requires that the yields in both the numerator and the denominator must have identical units, including the same chemical species.

EXERCISES

EXERCISE 3-1

Answer the following questions.

1. In a balanced chemical equation, the substances on the left side of the reaction arrow are the _____.

2. A chemical equation must be balanced to follow the _____.

3. The number of particles in a mole is known as _____, which is numerically to _____/mol.

4. The mass, in grams, of a mole of a substance is the _____.

5. If the atomic weight of carbon is 12 amu/atom, and the atomic weight of oxygen is 16 amu/atom, what is the molar mass of carbon monoxide, CO?

6. The substance that controls how far a reaction will proceed is the _____.

7. The amount of product calculated to form in a reaction is the _____.

8. The simplest whole-number ratio of the elements present in a compound is the _____.

9. Convert each of the following to empirical formulas.
    a. hydrogen peroxide, H_2O_2
    b. glucose, $C_6H_{12}O_6$
    c. ethane, C_2H_6
    d. nitric acid, HNO_3
    e. ammonium peroxydisulfate, $(NH_4)_2S_2O_8$

10. The analysis of a gas sample found 21.4 g of sulfur and 50.8 g of fluorine. Determine the empirical formula of this compound.

11. Analysis of a dark brown solid showed 69.6% manganese, Mn, and 30.4% oxygen, O. Determine the empirical formula of the compound. (Hint: assume you have 100 g of sample. This means that 69.6% Mn is 69.6 g of Mn and 30.4% O is 30.4 g of O.)

12. A gaseous sample with a molar mass of approximately 254 g/mol was found to consist of 25.2% sulfur, S, and 74.8% fluorine, F. What is the molecular formula of this sample?

EXERCISE 3-2

Balance the following chemical equations.

1. _____ $N_2(g)$ + _____ $O_2(g)$ → _____ $N_2O_5(s)$

2. _____ $CoCl_2(s)$ + _____ $ClF_3(g)$ → _____ $CoF_3(s)$ + _____ $Cl_2(g)$

3. _____ $La(OH)_3(s)$ + _____ $H_2C_2O_4(s)$ → _____ $La_2(C_2O_4)_3(s)$ + _____ $H_2O(l)$

4. _____ $C_4H_{10}(g)$ + _____ $O_2(g)$ → _____ $CO_2(g)$ + _____ $H_2O(g)$

5. Ammonia gas, NH_3, reacts with fluorine gas, F_2, to form nitrogen gas, N_2, and gaseous hydrogen fluoride, HF.

EXERCISE 3-3

You are given 3.00 mol of the second reactant in each of the reactions in Exercise 3-2. How many moles of the first (only) product will form in each case?

EXERCISE 3-4

Balance the following chemical equations.

1. _____ $Br_2(l)$ + _____ $O_2(g)$ → _____ $Br_2O_3(s)$

2. _____ $PF_3(l)$ + _____ $H_2O(l)$ → _____ $H_3PO_3(l)$ + _____ $HF(l)$

3. _____ $Zn(OH)_2(s)$ + _____ $H_3VO_4(s)$ → _____ $Zn_3(VO_4)_2(s)$ +

_____ $H_2O(l)$

4. _____ $C_6H_{14}(l)$ + _____ $O_2(g)$ → _____ $CO_2(g)$ + _____ $H_2O(l)$

5. Methane gas, CH_4, reacts with chlorine gas, Cl_2, to produce liquid carbon tetrachloride, CCl_4, and gaseous hydrogen chloride, HCl.

EXERCISE 3-5

You are given 3.00 mol of each reactant in each of the reactions in Exercise 3-4. Which of the reactants is the limiting reactant?

EXERCISE 3-6

You are given 3.00 mol of each reactant in each of the reactions in Exercise 3-4. What is the percent yield if 10.0 g of the first product forms?

EXERCISE 3-7

Determine the molar mass of each of the following to two decimal places.

1. H_2O

2. CO_2

3. HNO_3

4. Na_2SO_4

5. $(NH_4)_3PO_4$

EXERCISE 3-8

How many moles of material are present in 100.00 g of each of the following?

1. H_2O

2. CO_2

3. HNO_3

4. Na_2SO_4

5. $(NH_4)_3PO_4$

EXERCISE 3-9

How many grams are present in 2.50 mol of each of the following?

1. H_2O

2. CO_2

3. HNO_3

4. Na_2SO_4

5. $(NH_4)_3PO_4$

Flashcard
App

4 Aqueous Solutions

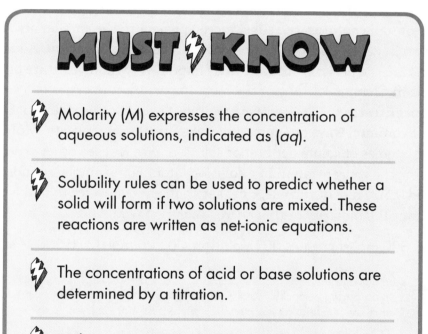

MUST KNOW

⚡ Molarity (M) expresses the concentration of aqueous solutions, indicated as (*aq*).

⚡ Solubility rules can be used to predict whether a solid will form if two solutions are mixed. These reactions are written as net-ionic equations.

⚡ The concentrations of acid or base solutions are determined by a titration.

⚡ Redox reactions involve the transfer of electrons.

In this chapter we are going to look at reactions in aqueous solutions, including acid-base titrations. We will also examine a set of solubility rules that can be used to predict whether precipitation will take place when two solutions are mixed. Let's get going.

Molarity (M)

A **solution** is a homogeneous (the same throughout) mixture composed of a **solvent**, the part of the solution that is present in the greater amount, and one or more **solutes**, the substance(s) present in the smaller amount. A solution in which water is the solvent is an **aqueous solution**. Aqueous solutions are the primary focus of this chapter. We will investigate other solutions in Chapter 12. One of the most important properties of a solution is its **concentration**—the relative amount of solute in the solution. One of the most common ways of expressing concentration is molarity. **Molarity** (M) is the moles of solute per liter of solution. (We will see other ways of expressing the concentration of a solution later.) For example, suppose you dissolved 0.500 mol of sucrose, cane sugar, in enough water to make 1.00 L of solution. The molarity of that sucrose solution would be:

$$0.500 \text{ mol sucrose}/1.00 \text{ L solution} = 0.500 \text{ mol/L} = 0.500 \text{ } M \text{ sucrose}$$

 IRL In preparing a specific molar solution, you should dissolve the required amount of solute in a little solvent and then dilute (add more solvent) to the required volume.

Solubility and Precipitation

Many of the reactions that you will study occur in aqueous solution. Water readily dissolves many ionic compounds as well as some covalent compounds. Ionic compounds that dissolve in water (dissociate) form

electrolyte solutions—solutions that conduct electrical current due to the presence of ions. We may classify electrolytes as either strong or weak. **Strong electrolytes** dissociate (break apart or ionize) completely in solution, while **weak electrolytes** only partially dissociate. Even though many ionic compounds dissolve in water, many do not. If the attraction of the oppositely charged ions in the solid is greater than the attraction of the water molecules to the ions, then the salt will not dissolve to an appreciable amount.

Compounds like alcohols and sucrose are **nonelectrolytes**—substances that do not conduct an electrical current when dissolved in water. However, certain covalent compounds, like acids, will **ionize** in water, that is, form ions:

$$HCl(aq) \rightarrow H^+(aq) + Cl^-(aq)$$

Precipitation reactions involve the formation of an insoluble compound, a **precipitate**, from the mixing of two aqueous solutions containing soluble compounds. To predict if precipitation will occur upon the mixing of two solutions, you **must know** and be able to apply the following solubility rules. You should apply these rules to all combinations of cations with anions in each of the mixed solutions. Be sure, however, that you do not try to break apart molecular species such as organic compounds.

Very few molecular compounds, other than acids, are electrolytes. Molecular compounds, as introduced in Chapter 2, are generally compounds composed entirely of nonmetals.

General solubility rules for the commonly encountered ions:

- All salts of Group 1A (Na$^+$, K$^+$, etc.) and the ammonium ion (NH$_4^+$) are *soluble*.

- All salts containing nitrate (NO$_3^-$), acetate (C$_2$H$_3$O$_2^-$), and perchlorate (ClO$_4^-$) are *soluble*.

- All chlorides (Cl$^-$), bromides (Br$^-$), and iodides (I$^-$) are *soluble except* those of Cu$^+$, Ag$^+$, Pb^{2+}, and Hg$_2^{2+}$.

■ All salts containing sulfate (SO_4^{2-}) are *soluble except* those of Pb^{2+}, Ca^{2+}, Sr^{2+}, and Ba^{2+}.

Salts containing the following ions are normally **insoluble**:

■ Most carbonates (CO_3^{2-}) and phosphates (PO_4^{3-}) are *insoluble except* those of Group 1A and the ammonium ion.

■ Most sulfides (S^{2-}) are *insoluble except* those of Groups 1A and 2A and the ammonium ion.

■ Most hydroxides (OH^-) are *insoluble except* those of Group 1A, calcium, strontium, and barium. (Later we will see that these are the strong bases.)

■ Most oxides (O^{2-}) are *insoluble except* for those of Groups 1A and 2A, which react with water. (These will produce strong bases.)

Let's consider the following problem as an example of how you might apply these rules.

> **BTW**
>
> You **must know** the solubility rules to predict whether precipitation will occur. You should only apply the solubility rules to combinations of ions.

EXAMPLE

▶ Suppose a solution of lead(II) nitrate is mixed with a solution of sodium iodide. Predict what will happen.

▶ Write the formulas for the reactants:

$$Pb(NO_3)_2(aq) + KI(aq)$$

▶ We need to know which ions are available. To do this, it will be helpful to break apart any ionic reactants into their constituent ions:

$$Pb^{2+}(aq) + 2\ NO_3^-(aq) + K^+(aq) + I^-(aq)$$

▸ Next, to predict possible products, we must consider all possible combinations of cations with anions. Predict the possible products by combining the cation of one reactant with the anion of the other and vice versa. In this case, we get:

You will never get a compound containing only cations or only anions. The total charge for a compound must be zero.

$$PbI_2 + KNO_3$$

▸ Apply the solubility rules to the two possible products:

$$PbI_2(s) \rightarrow \text{insoluble, therefore a precipitate will form}$$

$$KNO_3(aq) \rightarrow \text{soluble, no precipitate will form}$$

▸ Write the chemical equation by adding the predicted compounds to the product side, and balance the resultant equation:

$$Pb(NO_3)_2(aq) + 2\ KI(aq) \rightarrow PbI_2(s) + 2\ KNO_3(aq)$$

▸ If neither of the possible products were insoluble, then no precipitation reaction would occur. The solution would be a mixture of all the ions. Situations where "no reaction" occurs are often labeled **NR**.

Acids, Bases, and Neutralization

Acids and bases are extremely common substances, as are their reactions with each other. At the macroscopic level, acids taste sour (i.e., like lemon juice) and react with bases to yield salts. Bases taste bitter (i.e., like tonic water) and react with acids to form salts.

At the microscopic level, the Arrhenius theory defines **acids** as substances that when dissolved in water yield the hydronium ion (H_3O^+) or $H^+(aq)$. **Bases** are defined as substances that when dissolved in water yield the hydroxide ion (OH^-). Acids and bases may be **strong** (as in strong electrolytes), dissociating completely in water, or **weak** (as in weak electrolytes), partially dissociating in water. (We will see the more useful Brønsted–Lowry definitions of acids and bases in Chapter 15.) You **must know** the strong acids.

Strong acids include:

- hydrochloric acid, HCl

- hydrobromic acid, HBr

- hydroiodic acid, HI

- nitric acid, HNO_3

- sulfuric acid, H_2SO_4

- chloric acid, $HClO_3$

- perchloric acid, $HClO_4$

Strong bases include:

- alkali metal (Group 1A or Group 1) hydroxides (LiOH, NaOH, and so on)

- calcium, strontium, and barium hydroxides

Assume, unless told otherwise, that acids and bases not on the preceding lists are weak.

Certain oxides can have acidic or basic properties. Many oxides of metals that have a +1 or +2 charge are basic oxides because they will react with water to form a basic solution:

Hydrofluoric acid, HF, is not a strong acid.

$$Na_2O(s) + H_2O(l) \rightarrow 2\ NaOH(aq)$$

Many nonmetal oxides are acidic oxides because they react with water to form an acidic solution:

$$CO_2(g) + H_2O(l) \rightarrow H_2CO_3(aq)$$

$H_2CO_3(aq)$ is named carbonic acid and is one of the reasons that most carbonated beverages are slightly acidic. It is also the reason that soft drinks have fizz, because this carbonic acid can easily revert to carbon dioxide and water.

In general, acids react with bases to form water and a salt. The salt will depend upon what acid and base are used:

$$HCl(aq) + NaOH(aq) \rightarrow H_2O(l) + NaCl(aq)$$

$$HNO_3(aq) + KOH(aq) \rightarrow H_2O(l) + KNO_3(aq)$$

Reactions of this type are **neutralization reactions**.

Acids will also react with carbonates and acid carbonates (hydrogen carbonates, HCO_3^-) to form carbonic acid, which then decomposes to carbon dioxide and water:

BTW
The salt will always contain a cation from the base and an anion from the acid.

$$2\ HCl(aq) + Na_2CO_3(aq) \rightarrow 2\ NaCl(aq) + H_2CO_3(aq)$$
$$\downarrow$$
$$CO_2(g) + H_2O(l)$$

Oxidation-Reduction

Oxidation-reduction reactions, commonly called **redox** reactions, are an extremely important category of reactions. Redox reactions include combustion, corrosion, respiration, photosynthesis, and the reactions occurring in batteries.

Redox is a term that stands for **red**uction and **ox**idation. **Reduction** is the gain of electrons and **oxidation** is the loss of electrons. In these reactions, the number of electrons gained must be identical to the number of electrons lost. For example, suppose a piece of zinc metal is placed in a solution containing the Cu^{2+} cation. Very quickly, some reddish solid forms on the surface of the zinc metal. That substance is copper metal. At the molecular level, the zinc metal is losing electrons to form Zn^{2+} cations and the Cu^{2+} ions are gaining electrons to form copper metal. We can represent these two processes as:

$$Zn(s) \rightarrow Zn^{2+}(aq) + 2\ e^- \quad \text{oxidation}$$
$$Cu^{2+}(aq) + 2\ e^- \rightarrow Cu(s) \quad\quad\quad \text{reduction}$$

The electrons that are being lost by the zinc metal are the same electrons that are being gained by the copper(II) ion. The zinc metal is being oxidized and the copper(II) ion is being reduced.

Something must cause the oxidation (taking the electrons), and that substance is the **oxidizing agent** (the reactant undergoing reduction). In the preceding example, the oxidizing agent is the Cu^{2+} ion. The reactant undergoing oxidation is the **reducing agent** because it is furnishing the electrons that are being used in the reduction half-reaction. Zinc metal is the reducing agent above. The two half-reactions, oxidation and reduction, can be added together to give you the overall redox reaction. When doing this, the electrons must cancel—that is, there must be the same number of electrons lost as electrons gained:

$$Zn(s) + Cu^{2+}(aq) + \cancel{2\ e^-} \rightarrow Zn^{2+}(aq) + \cancel{2\ e^-} + Cu(s)$$

or

$$Zn(s) + Cu^{2+}(aq) \rightarrow Zn^{2+}(aq) + Cu(s)$$

In these redox reactions, there is a simultaneous loss and gain of electrons. In the oxidation reaction part of the reaction (oxidation half-reaction), electrons are being lost, but in the reduction half-reaction, those very same electrons are being gained. Therefore, in redox reactions there is an exchange of electrons, as reactants become products. This electron exchange may be direct, as when copper metal plates out on a piece of zinc, or it may be indirect, as in an electrochemical cell (battery).

Another way to determine what is undergoing oxidation and what is undergoing reduction is by looking at the change in oxidation numbers of the reactant species. Oxidation occurs when there is an increase in oxidation number. In the preceding example, the Zn metal went from an oxidation state of 0 to +2. Reduction occurs when there is a decrease in oxidation number. Cu^{2+} went from an oxidation state of +2 to 0. To determine if a reaction is a redox reaction, determine the oxidation numbers of each element in the reaction. If at least one element changes oxidation number, it is a redox reaction.

> **BTW**
> Oxidizing and reducing agents are reactants, not products.

We may predict many redox reactions of metals by using an **activity series**. An activity series lists reactions showing how various metals and hydrogen oxidize in aqueous solution. Elements at the top of the series are more reactive (active) than elements below. A reaction occurs when an element (not an ion) interacts with a cation of an element lower in the series. The more active elements have a stronger tendency to oxidize than the less active elements. The less active elements tend to reduce instead of oxidize. The reduction reactions are the reverse of the oxidation reactions given in the activity series table on the next page. This is an abbreviated table. Refer to the internet for a more complete table.

> **BTW**
> These types of reactions are sometimes called displacement reactions in which one atom is displacing another atom from a compound.

We may use the activity series to predict certain types of redox reactions. For example, suppose you wanted to write the equation between magnesium metal and hydrochloric acid:

$$Mg(s) + HCl(aq) \rightarrow$$

Activity Series of Some Common Metals

Hydrochloric acid is a strong acid (strong electrolyte). Therefore, the species present would be Mg(s), H$^+$(aq), and Cl$^-$(aq). Locate the element (Mg) and the cation (H$^+$) in the activity series.

Li(s)	→	Li$^+$(aq)	+	e$^-$
Mg(s)	→	Mg^{2+}(aq)	+	2 e$^-$
Al(s)	→	Al^{3+}(aq)	+	3 e$^-$
Zn(s)	→	Zn^{2+}(aq)	+	2 e$^-$
Fe(s)	→	Fe^{2+}(aq)	+	2 e$^-$
Ni(s)	→	Ni^{2+}(aq)	+	2 e$^-$
H$_2$(g)	→	2 H$^+$(aq)	+	2 e$^-$
Cu(s)	→	Cu^{2+}(aq)	+	2 e$^-$

$$Mg(s) \rightarrow Mg^{2+}(aq) + 2\ e^-$$
$$H_2(g) \rightarrow 2\ H^+(aq) + 2\ e^-$$

For there to be a reaction between the two, the *element* must be above the *cation* in the activity series. Since Mg is higher on the table, it will oxidize (react as shown in the table). (*Note*: only Mg, not Mg^{2+}, undergoes oxidation.) If a substance undergoes oxidation, then something *must* also undergo reduction. In this problem, the only option for a reduction species is the H$^+$. The reduction is the reverse of the equation in the table.

BTW
*The element in the activity series **never** has a charge.*

Net Ionic Equations

There are several ways to represent reactions in water. Suppose, for example, that we were writing an equation to describe the mixing of a lead(II) nitrate solution with a sodium sulfate solution and showing the resulting formation of lead(II) sulfate solid. One type of equation is the **molecular equation** in which both the reactants and products are in the undissociated form:

$$Pb(NO_3)_2(aq) + Na_2SO_4(aq) \rightarrow PbSO_4(s) + 2\ NaNO_3(aq)$$

In a molecular equation, we pretend that everything is a molecule (a nonelectrolyte). Molecular equations are quite useful when doing reaction stoichiometry problems.

Showing the strong electrolytes in the form of ions yields the **ionic equation** (sometimes called the total or overall ionic equation). The strong electrolytes are any strong acid, strong base, or water soluble (according to the solubility rules) ionic compound. In this example, the ionic equation is:

$$Pb^{2+}(aq) + 2\ NO_3^-(aq) + 2\ Na^+(aq) + SO_4^{2-}(aq) \rightarrow$$
$$PbSO_4(s) + 2\ Na^+(aq) + 2\ NO_3^-(aq)$$

Writing the equation in the ionic form shows clearly which species are really reacting and which are not. In the preceding example, the Na^+ and NO_3^- appear on both sides of the equation. They do not react but are simply there to maintain electrical neutrality of the solution. Ions like this, which are not actually involved in the chemical reaction, are **spectator ions**.

We write the **net ionic equation** by dropping out the spectator ions and showing only those chemical species that are involved in the chemical reaction:

$$Pb^{2+}(aq) + \cancel{2\ NO_3^-(aq)} + \cancel{2\ Na^+(aq)} + SO_4^{2-}(aq) \rightarrow$$
$$PbSO_4(s) + \cancel{2\ Na^+(aq)} + \cancel{2\ NO_3^-(aq)}$$

leaving

$$Pb^{2+}(aq) + SO_4^{2-}(aq) \rightarrow PbSO_4(s)$$

This net ionic equation focuses on the reactants only. It indicates that an aqueous solution containing Pb^{2+} (any solution, not just $Pb(NO_3)_2(aq)$) will react with any solution containing the sulfate ion to form insoluble lead(II) sulfate. If the net ionic equation form is used, we don't need to know the spectator ions involved. In most cases, this is not a problem.

Let's review with some more examples. We will treat these as a group to give additional practice on each step before moving on to the next step. These multiple examples let you practice each step. Do each step on all the examples before moving to the next step.

▶ Let's begin with reactants:

$$Cr(OH)_3(s) + HClO_4(aq) \rightarrow$$
$$KCl(aq) + (NH_4)_3PO_4(aq) \rightarrow$$
$$Na_2CO_3(aq) + HCl(aq) \rightarrow$$
$$Mg(NO_3)_2(aq) + Ca(OH)_2(aq) \rightarrow$$
$$HNO_3(aq) + Ba(OH)_2(aq) \rightarrow$$
$$LiF(aq) + H_2SO_4(aq) \rightarrow$$

▶ First, we'll predict the products. To predict the products, we need to know what ions are available. This will require an examination of each compound to determine what ions are present. We will write this information below each of the equations:

$$Cr(OH)_3(s) + HClO_4(aq) \rightarrow$$
Cr^{3+} and OH^- with H^+ and ClO_4^-

$$KCl(aq) + (NH_4)_3PO_4(aq) \rightarrow$$
K^+ and Cl^- with NH_4^+ and PO_4^{3-}

$$Na_2CO_3(aq) + HCl(aq) \rightarrow$$
Na^+ and CO_3^{2-} with H^+ and Cl^-

$$Mg(NO_3)_2(aq) + Ca(OH)_2(aq) \rightarrow$$
Mg^{2+} and NO_3^- with Ca^{2+} and OH^-

$$HNO_3(aq) + Ba(OH)_2(aq) \rightarrow$$
H^+ and NO_3^- with Ba^{2+} and OH^-

$$LiF(aq) + H_2SO_4(aq) \rightarrow$$
Li^+ and F^- with H^+ and SO_4^{2-}

BTW

If the phases (s, l, g, aq) are not present, you may wish to add these terms, if possible, because they are useful reminders.

BTW

*Each of these compounds is treated as a combination of cations (positive ions) and anions (negative ions). You will **never** have a compound of only cations or only anions.*

▶ The number of each type of ion is not important at this stage, and we do not care, at this point, which are soluble, strong electrolytes, or weak electrolytes.

$$Cr(OH)_3(s) + HClO_4(aq) \rightarrow Cr(ClO_4)_3 + H_2O$$
$$(Cr^{3+} \text{ and } OH^- \text{ with } H^+ \text{ and } ClO_4{}^-) \rightarrow (Cr^{3+} \text{ and } ClO_4{}^- \text{ with } H^+ \text{ and } OH^-)$$

$$KCl(aq) + (NH_4)_3PO_4(aq) \rightarrow K_3PO_4 + NH_4Cl$$
$$(K^+ \text{ and } Cl^- \text{ with } NH_4{}^+ \text{ and } PO_4{}^{3-}) \rightarrow (K^+ \text{ and } PO_4{}^{3-} \text{ with } NH_4{}^+ \text{ and } Cl^-)$$

$$Na_2CO_3(aq) + HCl(aq) \rightarrow NaCl + H_2CO_3$$
$$(Na^+ \text{ and } CO_3{}^{2-} \text{ with } H^+ \text{ and } Cl^-) \rightarrow (Na^+ \text{ and } Cl^- \text{ with } H^+ \text{ and } CO_3{}^{2-})$$

$$Mg(NO_3)_2(aq) + Ca(OH)_2(aq) \rightarrow Mg(OH)_2 + Ca(NO_3)_2$$
$$(Mg^{2+} \text{ and } NO_3{}^- \text{ with } Ca^{2+} \text{ and } OH^-) \rightarrow (Mg^{2+} \text{ and } OH^- \text{ with } Ca^{2+} \text{ and } NO_3{}^-)$$

$$HNO_3(aq) + Ba(OH)_2(aq) \rightarrow Ba(NO_3)_2 + H_2O$$
$$(H^+ \text{ and } NO_3{}^- \text{ with } Ba^{2+} \text{ and } OH^-) \rightarrow (Ba^{2+} \text{ and } NO_3{}^- \text{ with } H^+ \text{ and } OH^-)$$

$$LiF(aq) + H_2SO_4(aq) \rightarrow Li_2SO_4 + HF$$
$$(Li^+ \text{ and } F^- \text{ with } H^+ \text{ and } SO_4{}^{2-}) \rightarrow (Li^+ \text{ and } SO_4{}^{2-} \text{ with } H^+ \text{ and } F^-)$$

▶ To predict the products, we combine each cation with the anion of the other compound. We do this combination in a manner to give an overall charge of zero.

▶ There is one complication in these results. The compound carbonic acid, H_2CO_3, is unstable in water. It will rapidly decompose to gaseous carbon dioxide and liquid water. Therefore, we should replace H_2CO_3 with $CO_2 + H_2O$; however, this may be postponed until later if you wish.

▶ We can now balance each reaction by inspection:

$$Cr(OH)_3(s) + 3\ HClO_4(aq) \rightarrow Cr(ClO_4)_3 + 3\ H_2O$$
$$3\ KCl(aq) + (NH_4)_3PO_4(aq) \rightarrow K_3PO_4 + 3\ NH_4Cl$$
$$Na_2CO_3(aq) + 2\ HCl(aq) \rightarrow 2\ NaCl + H_2CO_3$$
$$Mg(NO_3)_2(aq) + Ca(OH)_2(aq) \rightarrow Mg(OH)_2 + Ca(NO_3)_2$$
$$2\ HNO_3(aq) + Ba(OH)_2(aq) \rightarrow Ba(NO_3)_2 + 2\ H_2O$$
$$2\ LiF(aq) + H_2SO_4(aq) \rightarrow Li_2SO_4 + 2\ HF$$

▶ These are the balanced molecular equations. To simplify the equations, we have omitted the phase designations temporarily. They will be added in the final answers. In general, you should use them.

▶ We now need to separate all the strong electrolytes into their component ions. We may begin with any category of strong electrolyte. In these examples, we will begin with the strong acids. Below each of the strong acids, we will write the separated ions:

$$Cr(OH)_3(s) + 3\ HClO_4(aq) \rightarrow Cr(ClO_4)_3 + 3\ H_2O$$
$$3\ H^+ + 3\ ClO_4^-$$
$$3\ KCl(aq) + (NH_4)_3PO_4(aq) \rightarrow K_3PO_4 + 3\ NH_4Cl$$
$$Na_2CO_3(aq) + 2\ HCl(aq) \rightarrow 2\ NaCl + CO_2 + H_2O$$
$$2\ H^+ + 2\ Cl^-$$
$$Mg(NO_3)_2(aq) + Ca(OH)_2(aq) \rightarrow Mg(OH)_2 + Ca(NO_3)_2$$
$$2\ HNO_3(aq) + Ba(OH)_2(aq) \rightarrow Ba(NO_3)_2 + 2\ H_2O$$
$$2\ H^+ + 2\ NO_3^-$$
$$2\ LiF(aq) + H_2SO_4(aq) \rightarrow Li_2SO_4 + 2\ HF$$
$$2\ H^+ + SO_4^{2-} \qquad —$$

▶ Notice that we have distributed the coefficient to each of the ions present. In addition, we did not separate the one remaining acid, HF, because it is a weak acid. There is a dash below the HF to remind us not to separate this compound.

▶ We will now separate all the strong bases and place a dash below any base that is not strong:

$$Cr(OH)_3(s) + 3\ HClO_4(aq) \rightarrow Cr(ClO_4)_3 + 3\ H_2O$$
$$\quad\text{—} \qquad 3\ H^+ + 3\ ClO_4^-$$

$$3\ KCl(aq) + (NH_4)_3PO_4(aq) \rightarrow K_3PO_4 + 3\ NH_4Cl$$

$$Na_2CO_3(aq) + 2\ HCl(aq) \rightarrow 2\ NaCl + CO_2 + H_2O$$
$$\qquad 2\ H^+ + 2\ Cl^-$$

$$Mg(NO_3)_2(aq) + Ca(OH)_2(aq) \rightarrow Mg(OH)_2 + Ca(NO_3)_2$$
$$\qquad Ca^{2+} + 2\ OH^- \qquad \text{—}$$

$$2\ HNO_3(aq) + Ba(OH)_2(aq) \rightarrow Ba(NO_3)_2 + 2\ H_2O$$
$$2\ H^+ + 2\ NO_3^-\ Ba^{2+} + 2\ OH^-$$

$$2\ LiF(aq) + H_2SO_4(aq) \rightarrow Li_2SO_4 + 2\ HF$$
$$\qquad 2\ H^+ + SO_4^{2-} \qquad \text{—}$$

▶ As a reminder, we will place dashes below each of the molecular compounds (CO_2 and H_2O) because they are nonelectrolytes, and we do not want to separate them by mistake. You can perform this step at any time in this process:

$$Cr(OH)_3(s) + 3\ HClO_4(aq) \rightarrow Cr(ClO_4)_3 + 3\ H_2O$$
$$\quad\text{—} \qquad 3\ H^+ + 3\ ClO_4^- \qquad \text{—}$$

$$3\ KCl(aq) + (NH_4)_3PO_4(aq) \rightarrow K_3PO_4 + 3\ NH_4Cl$$

$$Na_2CO_3(aq) + 2\ HCl(aq) \rightarrow 2\ NaCl + CO_2 + H_2O$$
$$2\ H^+ + 2\ Cl^- \qquad\qquad \text{—} \quad \text{—}$$

$$Mg(NO_3)_2(aq) + Ca(OH)_2(aq) \rightarrow Mg(OH)_2 + Ca(NO_3)_2$$
$$\qquad Ca^{2+} + 2\ OH^- \qquad \text{—}$$

$$2\ HNO_3(aq) + Ba(OH)_2(aq) \rightarrow Ba(NO_3)_2 + 2\ H_2O$$
$$2\ H^+ + 2\ NO_3^-\ Ba^{2+} + 2\ OH^- \qquad\qquad \text{—}$$

$$2\ LiF(aq) + H_2SO_4(aq) \rightarrow Li_2SO_4 + 2\ HF$$
$$\qquad 2\ H^+ + SO_4^{2-} \qquad \text{—}$$

▶ Finally, we will separate all soluble ionic compounds. We need only consider those compounds not already separated or that do not have a dash beneath the formula:

$$Cr(OH)_3(s) + 3\ HClO_4(aq) \rightarrow Cr(ClO_4)_3 + 3\ H_2O$$
$$\underline{\qquad} \qquad 3\ H^+ + 3\ ClO_4^- \rightarrow Cr^{3+} + 3\ ClO_4^- \qquad \underline{\qquad}$$

$$3\ KCl(aq) + (NH_4)_3PO_4(aq) \rightarrow K_3PO_4 + 3\ NH_4Cl$$
$$3\ K^+ + 3\ Cl^- \qquad 3\ NH_4^+ + PO_4^{3-} \rightarrow 3\ K^+ + PO_4^{3-} \quad 3\ NH_4^+ + 3\ Cl^-$$

$$Na_2CO_3(aq) + 2\ HCl(aq) \rightarrow 2\ NaCl + CO_2 + H_2O$$
$$2\ Na^+ + CO_3^{2-} \qquad 2\ H^+ + 2\ Cl^- \rightarrow 2\ Na^+ + 2\ Cl^- \qquad \underline{\qquad} \quad \underline{\qquad}$$

$$Mg(NO_3)_2(aq) + Ca(OH)_2(aq) \rightarrow Mg(OH)_2 + Ca(NO_3)_2$$
$$Mg^{2+} + 2\ NO_3^- \qquad Ca^{2+} + 2\ OH^- \rightarrow \qquad \underline{\qquad} \qquad Ca^{2+} + 2\ NO_3^-$$

$$2\ HNO_3(aq) + Ba(OH)_2(aq) \rightarrow Ba(NO_3)_2 + 2\ H_2O$$
$$2\ H^+ + 2\ NO_3^- \qquad Ba^{2+} + 2\ OH^- \rightarrow Ba^{2+} + 2\ NO_3^- \qquad \underline{\qquad}$$

$$2\ LiF(aq) + H_2SO_4(aq) \rightarrow Li_2SO_4 + 2\ HF$$
$$2\ Li^+ + 2\ F^- \qquad 2\ H^+ + SO_4^{2-} \rightarrow 2\ Li^+ + SO_4^{2-} \qquad \underline{\qquad}$$

▶ We can now write the complete or total ionic equation for each of the reactions. All we need to do for this step is to write the separated ions and non-ionized species on one line:

$$Cr(OH)_3(s) + 3\ H^+ + 3\ ClO_4^- \rightarrow Cr^{3+} + 3\ ClO_4^- + 3\ H_2O$$

$$3\ K^+ + 3\ Cl^- + 3\ NH_4^+ + PO_4^{3-} \rightarrow 3\ K^+ + PO_4^{3-} + 3\ NH_4^+ + 3\ Cl^-$$

$$2\ Na^+ + CO_3^{2-} + 2\ H^+ + 2\ Cl^- \rightarrow 2\ Na^+ + 2\ Cl^- + CO_2 + H_2O$$

$$Mg^{2+} + 2\ NO_3^- + Ca^{2+} + 2\ OH^- \rightarrow Mg(OH)_2 + Ca^{2+} + 2\ NO_3^-$$

$$2\ H^+ + 2\ NO_3^- + Ba^{2+} + 2\ OH^- \rightarrow Ba^{2+} + 2\ NO_3^- + 2\ H_2O$$

$$2\ Li^+ + 2\ F^- + 2\ H^+ + SO_4^{2-} \rightarrow 2\ Li^+ + SO_4^{2-} + 2\ HF$$

▶ All the separated ions should have "(aq)" after their formulas, as should the hydrofluoric acid. Water is a liquid (l), and carbon dioxide is a gas (g). The insoluble ionic materials, including those that are not strong bases, are solids (s). When we add this information, we have the ionic equations:

$$Cr(OH)_3(s) + 3\ H^+(aq) + 3\ ClO_4^-(aq) \rightarrow Cr^{3+}(aq) + 3\ ClO_4^-(aq) + 3\ H_2O(l)$$

$$3K^+(aq) + 3Cl^-(aq) + 3NH_4^+(aq) + PO_4^{3-}(aq) \rightarrow 3K^+(aq) + PO_4^{3-}(aq) + 3NH_4^+(aq) + 3Cl^-(aq)$$

$$2\ Na^+(aq) + CO_3^{2-}(aq) + 2\ H^+(aq) + 2\ Cl^-(aq) \rightarrow 2\ Na^+(aq) + 2\ Cl^-(aq) + CO_2(g) + H_2O(l)$$

$$Mg^{2+}(aq) + 2\ NO_3^-(aq) + Ca^{2+}(aq) + 2\ OH^-(aq) \rightarrow Mg(OH)_2(s) + Ca^{2+}(aq) + 2\ NO_3^-(aq)$$

$$2\ H^+(aq) + 2\ NO_3^-(aq) + Ba^{2+}(aq) + 2\ OH^-(aq) \rightarrow Ba^{2+}(aq) + 2\ NO_3^-(aq) + 2\ H_2O(l)$$

$$2\ Li^+(aq) + 2\ F^-(aq) + 2\ H^+(aq) + SO_4^{2-}(aq) \rightarrow 2\ Li^+(aq) + SO_4^{2-}(aq) + 2\ HF(aq)$$

▶ You can go back and add these phase designations to the original balanced molecular equations. The "(aq)" designation will go with each compound that you separated into ions. The resultant molecular equations are:

$$Cr(OH)_3(s) + 3\ HClO_4(aq) \rightarrow Cr(ClO_4)_3(aq) + 3\ H_2O(l)$$

$$3\ KCl(aq) + (NH_4)_3PO_4(aq) \rightarrow K_3PO_4(aq) + 3\ NH_4Cl(aq)$$

$$Na_2CO_3(aq) + 2\ HCl(aq) \rightarrow 2\ NaCl(aq) + CO_2(g) + H_2O(l)$$

$$Mg(NO_3)_2(aq) + Ca(OH)_2(aq) \rightarrow Mg(OH)_2(s) + Ca(NO_3)_2(aq)$$

$$2\ HNO_3(aq) + Ba(OH)_2(aq) \rightarrow Ba(NO_3)_2(aq) + 2\ H_2O(l)$$

$$2\ LiF(aq) + H_2SO_4(aq) \rightarrow Li_2SO_4(aq) + 2\ HF(aq)$$

▶ To get the net ionic equations, we must remove the spectator ions from the total ionic equations. We eliminate anything that appears identical on both sides of the reaction arrow. This cancellation gives:

$$Cr(OH)_3(s) + 3\ H^+(aq) + \cancel{3\ ClO_4^-(aq)} \rightarrow Cr^{3+}(aq) + \cancel{3\ ClO_4^-(aq)} +$$
$$3\ H_2O(l)$$

$$\cancel{3K^+(aq)} + \cancel{3Cl^-(aq)} + \cancel{3NH_4^+(aq)} + \cancel{PO_4^{3-}(aq)} \rightarrow \cancel{3K^+(aq)} + \cancel{PO_4^{3-}(aq)}$$
$$+ \cancel{3NH_4^+(aq)} + \cancel{3Cl^-(aq)}$$

$$\cancel{2\ Na^+(aq)} + CO_3^{2-}(aq) + 2\ H^+(aq) + \cancel{2\ Cl^-(aq)} \rightarrow \cancel{2\ Na^+(aq)} +$$
$$\cancel{2\ Cl^-(aq)} + CO_2(g) + H_2O(l)$$

$$Mg^{2+}(aq) + \cancel{2\ NO_3^-(aq)} + \cancel{Ca^{2+}(aq)} + 2\ OH^-(aq) \rightarrow Mg(OH)_2(s) +$$
$$\cancel{Ca^{2+}(aq)} + \cancel{2\ NO_3^-(aq)}$$

$$2\ H^+(aq) + \cancel{2\ NO_3^-(aq)} + \cancel{Ba^{2+}(aq)} + 2\ OH^-(aq) \rightarrow \cancel{Ba^{2+}(aq)} +$$
$$\cancel{2\ NO_3^-(aq)} + 2\ H_2O(l)$$

$$\cancel{2\ Li^+(aq)} + 2\ F^-(aq) + 2\ H^+(aq) + \cancel{SO_4^{2-}(aq)} \rightarrow \cancel{2\ Li^+(aq)} + \cancel{SO_4^{2-}(aq)}$$
$$+ 2\ HF(aq)$$

▶ You will notice that in the second reaction, all species cancel. This will require us to amend our molecular equation to:

> KCl(aq) + (NH$_4$)$_3$PO$_4$(aq) ➝ no reaction
> (NR is commonly used.)

BTW

When canceling, identical means **identical**. For example, Ca is not identical to Ca²⁺.

▶ We can now rewrite the ionic equations without the cancelled species to have the net ionic equations:

$$Cr(OH)_3(s) + 3\ H^+(aq) \rightarrow Cr^{3+}(aq) + 3\ H_2O(l)$$
No Reaction [KCl(aq) + (NH$_4$)$_3$PO$_4$(aq)]
$$CO_3^{2-}(aq) + 2\ H^+(aq) \rightarrow CO_2(g) + H_2O(l)$$
$$Mg^{2+}(aq) + 2\ OH^-(aq) \rightarrow Mg(OH)_2(s)$$
$$2\ H^+(aq) + 2\ OH^-(aq) \rightarrow 2\ H_2O(l)$$
$$2\ F^-(aq) + 2\ H^+(aq) \rightarrow 2\ HF(aq)$$

▶ The last two equations need further simplification by reducing the coefficients by the common factor of two. The final net ionic equations are:

$Cr(OH)_3(s) + 3\ H^+(aq) \rightarrow Cr^{3+}(aq) + 3\ H_2O(l)$

$[KCl(aq) + (NH_4)_3PO_4(aq)]$ No Reaction

$CO_3^{2-}(aq) + 2\ H^+(aq) \rightarrow CO_2(g) + H_2O(l)$

$Mg^{2+}(aq) + 2\ OH^-(aq) \rightarrow Mg(OH)_2(s)$

$H^+(aq) + OH^-(aq) \rightarrow H_2O(l)$

$F^-(aq) + H^+(aq) \rightarrow HF(aq)$

This method will work for most reactions. However, if a redox reaction is occurring, an additional consideration is necessary. At this point, the presence of hydrogen, H_2, or a metal indicates that redox is possible. We can determine if there will be redox by using the activity series.

EXAMPLE

▶ We will begin, as before, with a group of potential reactions:

$HCl(aq) + Ni(s) \rightarrow$

$CuBr_2(aq) + Zn(s) \rightarrow$

$Mg(s) + HC_2H_3O_2(aq) \rightarrow$

$Al_2(SO_4)_3(aq) + Fe(s) \rightarrow$

$Li(s) + H_2O(l) \rightarrow$

▶ We need to consider the constituent ions of the compounds.

$HCl(aq) + Ni(s) \rightarrow$
$H^+ + Cl^-$

$CuBr_2(aq) + Zn(s) \rightarrow$
$Cu^{2+} + Br^-$

$Mg(s) + HC_2H_3O_2(aq) \rightarrow$
$\qquad H^+ + C_2H_3O_2$

$$Al_2(SO_4)_3(aq) + Fe(s) \rightarrow$$
$$Al^{3+} + SO_4^{2-}$$
$$Li(s) + H_2O(l) \rightarrow$$
$$H^+ + OH^-$$

▸ As seen in the previous examples, we do not need to worry about the number of each type of ion present, only their identity. There is a difference in these examples that we did not see before, and that we are treating water as $H^+ + OH^-$.

▸ The anions, negative ions, are not of concern to us at this time. We only need to locate the elements and cations on the activity series table. Here is our abbreviated activity series table with the substances from the first reaction in boldface:

$$Li(s) \rightarrow Li^+(aq) + e^-$$
$$Mg(s) \rightarrow Mg^{2+}(aq) + 2\,e^-$$
$$Al(s) \rightarrow Al^{3+}(aq) + 3\,e^-$$
$$Zn(s) \rightarrow Zn^{2+}(aq) + 2\,e^-$$
$$Fe(s) \rightarrow Fe^{2+}(aq) + 2\,e^-$$
$$\textbf{Ni(s)} \rightarrow Ni^{2+}(aq) + 2\,e^-$$
$$H_2(g) \rightarrow 2\,\textbf{H}^+\textbf{(aq)} + 2\,e^-$$
$$Cu(s) \rightarrow Cu^{2+}(aq) + 2\,e^-$$

▸ The location of these species is very important. If there is to be a reaction, the substance on the left *must* be above the substance on the right. If this is not true, or if both substances are on the same side, there will be no reaction. If there is a reaction, the substance higher on the table will react as shown in the table, and the lower substance will react in the reverse direction. From the table, we need to copy the reactions (remembering to reverse the lower reaction). We need only the two equations from the table and to know their positions relative to each other.

$$HCl(aq) + Ni(s) \rightarrow$$
$$H^+ + Cl^-$$
$$Ni(s) \rightarrow Ni^{2+}(aq) + 2\ e^-$$
$$2\ H^+(aq) + 2\ e^- \rightarrow H_2(g)$$

▶ The other reactions give the following sets:

$$CuBr_2(aq) + Zn(s) \rightarrow$$
$$Cu^{2+} + Br^-$$
$$Zn(s) \rightarrow Zn^{2+}(aq) + 2\ e^-$$
$$Cu(s) \rightarrow Cu^{2+}(aq) + 2\ e^-$$

▶ There will be a reaction, so we need to reverse the second equation.

$$Mg(s) + HC_2H_3O_2(aq) \rightarrow$$
$$H^+ + C_2H_3O_2^-$$
$$Mg(s) \rightarrow Mg^{2+}(aq) + 2\ e^-$$
$$H_2(g) \rightarrow 2\ H^+(aq) + 2\ e^-$$

▶ There will be a reaction, so we need to reverse the second equation.

$$Al_2(SO_4)_3(aq) + Fe(s) \rightarrow$$
$$Al^{3+} + SO_4^{2-}$$
$$Al(s) \rightarrow Al^{3+}(aq) + 3\ e^-$$
$$Fe(s) \rightarrow Fe^{2+}(aq) + 2\ e^-$$

▶ There will be no reaction because the substance on the left is not higher on the activity series. (In this situation, it is useful to treat H_2O and H^+ as OH^-.)

$$Li(s) + H_2O(l) \rightarrow$$
$$H^+ + OH^-$$

$$Li(s) \rightarrow Li^+(aq) + e^-$$
$$H_2(g) \rightarrow 2\,H^+(aq) + 2\,e^-$$

▶ There will be a reaction, so we need to reverse the second equation. We now have the following set of equations:

$$Ni(s) \rightarrow Ni^{2+}(aq) + 2\,e^-$$
$$2\,H^+(aq) + 2\,e^- \rightarrow H_2(g)$$

$$Zn(s) \rightarrow Zn^{2+}(aq) + 2\,e^-$$
$$Cu^{2+}(aq) + 2\,e^- \rightarrow Cu(s)$$

$$Mg(s) \rightarrow Mg^{2+}(aq) + 2\,e^-$$
$$2\,H^+(aq) + 2\,e^- \rightarrow H_2(g)$$

$$Al_2(SO_4)_3(aq) + Fe(s) \rightarrow \text{no reaction}$$

$$Li(s) \rightarrow Li^+(aq) + e^-$$
$$2\,H^+(aq) + 2\,e^- \rightarrow H_2(g)$$

▶ We need to adjust these equations so the number of electrons in each pair of reactions matches. All the reactions, except the lithium equation, have two electrons. We need to multiply the lithium equation by 2, so its electrons will match those in the hydrogen reaction. We then add each pair of reactions and cancel the electrons:

$$Ni(s) + 2\,H^+(aq) + \cancel{2\,e^-} \rightarrow Ni^{2+}(aq) + \cancel{2\,e^-} + H_2(g)$$

$$Zn(s) + Cu^{2+}(aq) + \cancel{2\,e^-} \rightarrow Zn^{2+}(aq) + \cancel{2\,e^-} + Cu(s)$$

$$Mg(s) + 2\,H^+(aq) + \cancel{2\,e^-} \rightarrow Mg^{2+}(aq) + \cancel{2\,e^-} + H_2(g)$$

$$Al_2(SO_4)_3(aq) + Fe(s) \rightarrow \text{no reaction}$$

$$2\,Li(s) + 2\,H^+(aq) + \cancel{2\,e^-} \rightarrow 2\,Li^+(aq) + \cancel{2\,e^-} + H_2(g)$$

▶ Later, we will see that the first and second reactions are the net ionic equations for these reactions.

▶ To get to the molecular equation we must put the ions in the preceding reactions into their original compounds.

$$Ni(s) + 2\ HCl(aq) \rightarrow Ni^{2+}(aq) + H_2(g)$$

$$Zn(s) + CuBr_2(aq) \rightarrow Zn^{2+}(aq) + Cu(s)$$

$$Mg(s) + 2\ HC_2H_3O_2(aq) \rightarrow Mg^{2+}(aq) + H_2(g)$$

$$Al_2(SO_4)_3(aq) + Fe(s) \rightarrow no\ reaction$$

$$2\ Li(s) + 2\ H_2O(aq) \rightarrow 2\ Li^+(aq) + H_2(g)$$

▶ Completing the reactants requires the addition of the following ions: reaction 1, 2 Cl^-; reaction 2, 2 Br^-; reaction 3, 2 $C_2H_3O_2^-$; reaction 4, no ions; reaction 5, 2 OH^-. To balance the reactions, we add these ions to the other side of the reaction arrow and combine with the cations present on the other side. This gives the molecular equations:

$$Ni(s) + 2\ HCl(aq) \rightarrow NiCl_2(aq) + H_2(g)$$

$$Zn(s) + CuBr_2(aq) \rightarrow ZnBr_2(aq) + Cu(s)$$

$$Mg(s) + 2\ HC_2H_3O_2(aq) \rightarrow Mg(C_2H_3O_2)_2(aq) + H_2(g)$$

$$Al_2(SO_4)_3(aq) + Fe(s) \rightarrow no\ reaction$$

$$2\ Li(s) + 2\ H_2O(aq) \rightarrow 2\ LiOH(aq) + H_2(g)$$

▶ We can now follow the steps used previously to convert each of these to a total ionic equation:

$$Ni(s) + 2\ H^+(aq) + 2\ Cl^-(aq) \rightarrow Ni^{2+}(aq) + 2\ Cl^-(aq) + H_2(g)$$

$$Zn(s) + Cu^{2+}(aq) + 2\ Br^-(aq) \rightarrow Zn^{2+}(aq) + 2\ Br^-(aq) + Cu(s)$$

$$Mg(s) + 2\ HC_2H_3O_2(aq) \rightarrow Mg^{2+}(aq) + 2\ C_2H_3O_2^-(aq) + H_2(g)$$

$$Al_2(SO_4)_3(aq) + Fe(s) \rightarrow no\ reaction$$

$$2\ Li(s) + 2\ H_2O(aq) \rightarrow 2\ Li^+(aq) + 2\ OH^-(aq) + H_2(g)$$

▶ Finally, we eliminate the spectator ions to obtain the net ionic equations:

$$Ni(s) + 2\ H^+(aq) \rightarrow Ni^{2+}(aq) + H_2(g)$$

$$Zn(s) + Cu^{2+}(aq) \rightarrow Zn^{2+}(aq) + Cu(s)$$

$$Mg(s) + 2\ HC_2H_3O_2(aq) \rightarrow Mg^{2+}(aq) + 2\ C_2H_3O_2^-(aq) + H_2(g)$$

$$Al_2(SO_4)_3(aq) + Fe(s) \rightarrow \text{no reaction}$$

$$2\ Li(s) + 2\ H_2O(aq) \rightarrow 2\ Li^+(aq) + 2\ OH^-(aq) + H_2(g)$$

Titrations

A common laboratory application of acid–base reactions is titration. A **titration** is a laboratory procedure in which we use a solution of known concentration to determine some information (such as concentration and mass) about an unknown substance. A titration may involve any type of reaction–acid-base, redox, and so on. In this section, we will only consider acid-base titrations. The calculations for acid-base titrations are identical to those for any type of titration reaction. The key to any titration calculation will be moles.

In an acid-base titration you may either add acid to base or base to acid. This addition continues until there is some indication that the reaction is complete. Often a chemical known as an **indicator** will indicate the **endpoint** of a titration reaction; the endpoint is the experimental end of the titration. If we perform the experiment well, the endpoint should closely match the **equivalence point** of the titration; the equivalence point is the theoretical end of the titration reaction. All the calculations in this section assume accurate experimental determination of the endpoint and that this value is the same as the equivalence point.

We will work with the following acid-base titration reaction for the remainder of this section:

$$2 \text{ HC}_2\text{H}_3\text{O}_2(aq) + \text{Ca(OH)}_2(aq) \rightarrow \text{Ca(C}_2\text{H}_3\text{O}_2)_2(aq) + 2 \text{ H}_2\text{O}(l)$$

For example, we could use this reaction for determining the concentration of acetic acid, $\text{HC}_2\text{H}_3\text{O}_2$, in vinegar. A titration problem will give you information about one reactant and ask you for information about the other reactant. In most titration reactions, information about the products will not be necessary. You only need to consider the products when we need to balance the chemical equation.

If we want to use the concentration of acetic acid to find the concentration of calcium hydroxide, we will need this step (the numbers are the coefficients in the balanced chemical equation):

$$\left(\frac{1 \text{ mol Ca(OH)}_2}{2 \text{ mol HC}_2\text{H}_3\text{O}_2} \right)$$

If we want to use the calcium hydroxide to find out about the acetic acid, we will need this step:

$$\left(\frac{2 \text{ mol HC}_2\text{H}_3\text{O}_2}{1 \text{ mol Ca(OH)}_2} \right)$$

All possible titration problems for these substances will simply be adding steps before and/or after one of these two mole ratio terms. The problems will begin with the substance with the most information and proceed through one of these mole ratios to the substance with less information given.

BTW

In titration calculations, you must consider the reaction stoichiometry.

EXAMPLE

▶ Let's begin with this titration question: How many moles of calcium hydroxide are necessary to titrate 0.250 mol of acetic acid?

▶ We have information about the acetic acid, and we are seeking information about the calcium hydroxide. We will begin the problem with the acetic acid since we know more about it, and we will end the problem with the calcium hydroxide since we do not know much about this compound. One way to remember what you need to do is to copy the given information and the question to the balanced chemical equation, as shown here:

$$2\ HC_2H_3O_2(aq) + Ca(OH)_2(aq) \rightarrow Ca(C_2H_3O_2)_2(aq) + 2\ H_2O(l)$$

0.250 mol   ? mol

▶ When written this way, you begin with the number and work toward the question mark. We begin with the acetic acid and use the mole ratio with acetic acid in the denominator:

$$\left(0.250\ mol\ HC_2H_3O_2\right)\left(\frac{1\ mol\ Ca(OH)_2}{2\ mol\ HC_2H_3O_2}\right) = 0.125\ mol\ Ca(OH)_2$$

▶ This calculation leaves us with the appropriate number of significant figures. Note that the units of $HC_2H_3O_2$ have cancelled, leaving the desired units of mol $Ca(OH)_2$.

Let's now look at a variation of the typical titration problem.

EXAMPLE

▶ How would the problem be different if the titration question asked is: How many moles of calcium hydroxide are necessary to titrate 0.0500 L of a 0.100 M acetic acid solution?

▶ We can begin by adding our information to the balanced chemical equation:

$$2\ HC_2H_3O_2(aq) + Ca(OH)_2(aq) \rightarrow Ca(C_2H_3O_2)_2(aq) + 2\ H_2O(l)$$

0.0500 L   ? mol

0.100 M

▶ As in the preceding example, we will begin with the acetic acid and go through the same mole ratio to get to the calcium hydroxide. The difference in this problem is that we are beginning with liters and molarity instead of moles. We will need to add a step to our calculation to change liters and molarity to moles. Part of this step will use the definition of molarity, moles solute/liters of solution. For this problem, our setup will be:

$$(0.0500 \text{ L})\left(\frac{0.100 \text{ mol HC}_2\text{H}_3\text{O}_2}{\text{L}}\right)\left(\frac{1 \text{ mol Ca(OH)}_2}{2 \text{ mol HC}_2\text{H}_3\text{O}_2}\right)$$
$$= 2.50 \times 10^{-3} \text{ mol Ca(OH)}_2$$

> **BTW**
> *If the problem begins with milliliters instead of liters, you will need to convert the milliliters to liters.*

▶ This calculation leaves us with the appropriate number of significant figures and the correct units.

▶ The first part of this problem appears in numerous problems involving solutions. Moles are critical to all stoichiometry problems, so you will see this step repeatedly. This is so common that anytime you see a volume and a concentration of a solution, you should prepare to do this step.

> **BTW**
> *Molarity is a useful means of reporting concentrations. However, in most problems concerning molarity, you should use the definition (mol/L) instead of M.*

Now let us try an example needing additional information after the mole ratio step.

EXAMPLE

▶ How many grams of calcium hydroxide are necessary to titrate 0.200 mol of acetic acid?

▶ As usual, we begin by adding this information to the balanced chemical equation:

$$2 \text{ HC}_2\text{H}_3\text{O}_2(aq) + \text{Ca(OH)}_2(aq) \rightarrow \text{Ca(C}_2\text{H}_3\text{O}_2)_2(aq) + 2 \text{ H}_2\text{O}(l)$$
$$0.200 \text{ mol} \text{? g}$$

▶ The calculation will begin as in the first example in this section:

$$\left(0.200 \text{ mol HC}_2\text{H}_3\text{O}_2\right)\left(\frac{1 \text{ mol Ca(OH)}_2}{2 \text{ mol HC}_2\text{H}_3\text{O}_2}\right) =$$

▶ This will give us the moles of calcium hydroxide instead of the grams, thus we need to add another step. We have moles and we need grams; the molar mass relates these two quantities. We determine the molar mass by using the formula and the atomic weights found in a table such as the periodic table.

BTW

Anytime the problem deals with the moles and the mass of a substance, you will very, very likely need to know the molar mass (molecular weight) of the substance to complete the problem.

▶ The molar mass of calcium hydroxide is 74.10 g/mol. We can add the molar mass to our calculation to convert our moles to grams:

$$\left(0.200 \text{ mol HC}_2\text{H}_3\text{O}_2\right)\left(\frac{1 \text{ mol Ca(OH)}_2}{2 \text{ mol HC}_2\text{H}_3\text{O}_2}\right)\left(\frac{74.10 \text{ g Ca(OH)}_2}{1 \text{ mol Ca(OH)}_2}\right)$$

$$= 7.41 \text{ g Ca(OH)}_2$$

▶ This calculation leaves us with the appropriate number of significant figures and the correct units.

Now let's work one of the more typical titration problems.

EXAMPLE

▶ What is the concentration of a calcium hydroxide solution if 0.0250 L of a calcium hydroxide solution were necessary to titrate 0.0400 L of a 0.100 *M* acetic acid solution?

▶ We have more information, but this does not change our first step of transferring this information to the balanced chemical equation. You should recognize that by "concentration," the problem means molarity.

$$2 \text{ HC}_2\text{H}_3\text{O}_2(aq) + \text{Ca(OH)}_2(aq) \rightarrow \text{Ca(C}_2\text{H}_3\text{O}_2)_2(aq) + 2 \text{ H}_2\text{O}(l)$$

0.0400 L	? M
0.100 M	0.0250 L

▶ The calculation in this problem begins like the second example in this section:

$$\left(0.0400 \text{ L}\right)\left(\frac{0.100 \text{ mol HC}_2\text{H}_3\text{O}_2}{\text{L}}\right)\left(\frac{1 \text{ mol Ca(OH)}_2}{2 \text{ mol HC}_2\text{H}_3\text{O}_2}\right) =$$

▶ These steps give us the moles of calcium hydroxide. To get the molarity of calcium hydroxide, we use the definition of molarity, which tells us we need to divide these moles by the volume, in liters, of the calcium hydroxide solution. In order to finish the problem, we must add one more step:

$$\left(0.0400 \text{ } L\right)\left(\frac{0.100 \text{ mol HC}_2\text{H}_3\text{O}_2}{\text{L}}\right)\left(\frac{1 \text{ mol Ca(OH)}_2}{2 \text{ mol HC}_2\text{H}_3\text{O}_2}\right)\left(\frac{1}{0.0250 \text{ } L}\right)$$

$$= 0.0800 \text{ } M \text{ Ca(OH)}_2$$

▶ This calculation leaves us with the appropriate number of significant figures and the correct units.

It is possible to expand these examples to any titration problem, acid-base, redox, precipitation, and so on. Just remember that the key is the mole concept.

EXERCISES

EXERCISE 4-1

Answer the following questions.

1. True or False: A solution is a heterogeneous mixture.

2. True or False: Acids are electrolytes.

3. Which of the following is a molecular compound?
   a. NaCl
   b. KNO_3
   c. CH_3OH
   d. CsOH
   e. $(NH_4)_2SO_4$

4. Which of the following is not soluble in water?
   a. AgCl
   b. KCl
   c. NH_4Cl
   d. $FeCl_2$
   e. $FeCl_3$

5. Which of the following is insoluble in water?
   a. $(NH_4)_2SO_4$
   b. Na_2SO_4
   c. $MgSO_4$
   d. $ZnSO_4$
   e. $BaSO_4$

6. You are to mix two aqueous solutions to see if a precipitate will form. Which of the following combinations will not form a precipitate?
   a. $AgNO_3$ + KBr
   b. $Pb(NO_3)_2$ + KCl
   c.   NaOH + $FeCl_2$
   d. Na_3PO_4 + NH_4Cl
   e. $CaCl_2$ + K_2SO_4

7. Which of the following sets includes only strong acids?
   a.   HF, HCl, and HBr
   b. HNO_3, $HClO_3$, and $HClO_4$
   c.   HCl, HNO_3, and HNO_2
   d.   HCl, HI, and $HC_2H_3O_2$
   e.   HBr, HI, and H_2SO_3

8. Nonmetal oxides usually produce _____ when added to water.

9. Metal oxides often produce _____ when added to water.

10. A neutralization reaction requires an _____ and a _____.

11. A salt contains the cation from a _____ and the anion from an _____.

12. An oxidation-reduction reaction is often termed a _____ reaction.

13. To convert a molecular equation to an ionic equation, you must separate all _____ electrolytes.

14. To convert an ionic equation to a net ionic equation, you must eliminate all _____ ions.

15. The experimental end of a titration is the _____.

EXERCISE 4-2

Use the following reaction to answer the questions in this exercise.

$$\text{Mg}(s) + 2\,\text{Ag}^+(aq) \rightarrow \text{Mg}^{2+}(aq) + 2\,\text{Ag}(s)$$

1. Which element is undergoing oxidation in the reaction?

2. Which element is undergoing reduction in the reaction?

3. Which element is the oxidizing agent in the reaction?

4. Which element is the reducing agent in the reaction?

5. What is the charge of magnesium, Mg, on the reactant side of the reaction?

EXERCISE 4-3

Complete and balance each of the following equations, indicating no reaction, NR, where appropriate.

1. $\text{Ba(OH)}_2(aq) + \text{H}_2\text{SO}_4(aq) \rightarrow$

2. $\text{KOH}(aq) + \text{FeCl}_2(aq) \rightarrow$

3. $\text{NH}_4\text{NO}_3(aq) + \text{Na}_2\text{SO}_4(aq) \rightarrow$

4. $\text{HC}_2\text{H}_3\text{O}_2(aq) + \text{CaCO}_3(s) \rightarrow$

5. $\text{K}_3\text{PO}_4(aq) + \text{Ca(NO}_2)_2(aq) \rightarrow$

EXERCISE 4-4

Convert your answers from exercise 4-3 to ionic equations. (You may look at the answers to exercise 4-3 before doing this.)

EXERCISE 4-5

Convert your answers from exercise 4-4 to net ionic equations. (You may look at the answers to exercise 4-4 before doing this.)

EXERCISE 4-6

Using the activity series in this text, write net ionic equations for each of the following. Indicate no reaction, NR, where appropriate.

1. $Cu(NO_3)_2(aq) + Mg(s) \rightarrow$

2. $Al(s) + Ni(NO_3)_2(aq) \rightarrow$

3. $ZnCl_2(aq) + Cu(s) \rightarrow$

4. $Li(s) + H_2O(l) \rightarrow$

5. $HCl(aq) + Fe(s) \rightarrow$

EXERCISE 4-7

Use the following neutralization reaction for the questions in this exercise.

$$2\ HNO_2(aq) + Ca(OH)_2(aq) \rightarrow Ca(NO_2)_2(aq) + 2\ H_2O(l)$$

1. The titration of 25.00 mL of 0.1000 M $Ca(OH)_2$ required 30.00 mL of HNO_2 solution. What was the concentration of the HNO_2 solution?

2. The titration of 25.00 mL of 0.1000 M HNO_2 required 30.00 mL of $Ca(OH)_2$ solution. What was the concentration of the $Ca(OH)_2$ solution?

3. What volume of 0.1000 M HNO_2 solution is necessary to titrate 45.00 mL of 0.01000 M $Ca(OH)_2$?

4. What volume of 0.01500 M $Ca(OH)_2$ solution is necessary to titrate 40.00 mL of 0.2000 M HNO_2?

Flashcard App

5 Gases and Gas Laws

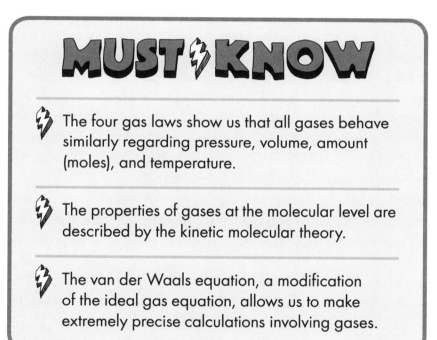

MUST ⚡ KNOW

⚡ The four gas laws show us that all gases behave similarly regarding pressure, volume, amount (moles), and temperature.

⚡ The properties of gases at the molecular level are described by the kinetic molecular theory.

⚡ The van der Waals equation, a modification of the ideal gas equation, allows us to make extremely precise calculations involving gases.

B efore we dive into the concepts of gases and gas laws, be sure that you know how to properly use your calculator, and if you need to, refer to Chapter 3 on the mole concept. It's especially true with gas law problems that the only way to master them is to work a lot of problems!

Gas Laws (P, V, n, and T)

Before we start describing the gas law relationships, we will need to describe the concept of pressure. When we use the word **pressure** with respect to gases, we may be referring to the pressure of a gas inside a container or we might be referring to atmospheric pressure, the pressure due to the weight of the atmosphere above us. The pressure at sea level is 1 atmosphere (atm). Commonly, the unit torr is used for pressure, where 1 torr = 1 mmHg (millimeters of mercury), so that atmospheric pressure at sea level equals 760 torr. The SI unit of pressure is the **pascal** (**Pa**), so that 1 atm = 760 mmHg = 760 torr = 1.01325×10^5 Pa (or 101.325 kPa). In the United States, pressure is often measured in inches (of mercury).

Dalton's law says that in a mixture of gases (A + B + C + . . .), the total pressure is simply the sum of the partial pressures (the pressures associated with each individual gas). Mathematically, Dalton's law looks like this:

$$P_{Total} = P_A + P_B + P_C + \ldots$$

If you know how many moles of each gas are in the mixture and the total pressure, you can calculate the partial pressure of each gas by multiplying the total pressure by the mole fraction of each gas:

$$P_A = (P_{Total})(X_A)$$

The term X_A refers to the mole fraction of gas A. The mole fraction of gas A is equal to the moles gas A/(total moles of gas in the mixture). The mole fraction is a concentration unit that we will see again in Chapter 12. (Mole fractions are not limited to gases.)

The gas laws relate the physical properties of pressure (P), volume (V), temperature (T), and amount (n = moles) to each other. If we keep the amount (number of moles of gas) constant, that is, no gas can get in or out, we can express the relationship between the other three by the **combined gas equation**:

BTW

In any gas law calculation, you must express the temperature in Kelvin.

$$\frac{P_1 V_1}{T_1} = \frac{P_2 V_2}{T_2}$$

The numbers 1 and 2 refer to the initial and final conditions of the gas, respectively.

EXAMPLE

▶ For example, suppose you have a balloon occupying a volume of 10.0 L at an internal pressure of 1.50 atm at 20.0°C. It is heated to 70.0°C and the pressure is determined to be 2.00 atm. We can calculate the new volume using the combined gas law.

▶ We will be solving the combined gas equation for V_2, so we will take the combined gas law and rearrange for V_2:

BTW

The temperatures must be expressed in Kelvin. 20.0°C = 293 K (K =°C + 273) and 70.0°C = 343 K.

$$V_2 = \frac{P_1 V_1 T_2}{T_1 P_2}$$

▶ Substituting in the values:

$$V_2 = \frac{(1.50 \text{ atm})(10.0 \text{ L})(343 \text{ K})}{(293 \text{ K})(2.00 \text{ atm})} = 8.77986 = 8.78 \text{ } L$$

IRL   A practical application of these gas laws is scuba diving. If a diver is at a deep depth and starts for the surface, the gas in her lungs starts to expand as the pressure on her body decreases. She must keep exhaling to reduce the volume of air in her lungs or her lungs could rupture.

In the combined gas equation, we held just the amount (moles) constant. If, however, we hold two quantities constant and look at the relationship between the other two, we can derive the other common gas laws shown as follows in the table of gas laws.

The Gas Laws

Name of Law	Held Constant	Variables	Relationship
combined gas law	amount (n)	pressure, volume, and temperature (K)	$(P_1 V_1)/T_1 = (P_2 V_2)/T_2$
Boyle's law	temperature (T) and amount (n)	volume and pressure	$P_1 V_1 = P_2 V_2$
Charles's law	pressure (P) and amount (n)	volume and temperature (K)	$V_1/T_1 = V_2/T_2$
Gay-Lussac's law	volume (V) and amount (n)	pressure and temperature (K)	$P_1/T_1 = P_2/T_2$
Avogadro's law	pressure (P) and temperature (T)	volume and amount	$V_1/n_1 = V_2/n_2$

It is possible to combine Avogadro's law and the combined gas law to produce the ideal gas equation, which incorporates the pressure, volume, temperature, and amount relationships of a gas. The **ideal gas equation** has the form of:

$$PV = nRT$$

The terms in this equation are:

> P = pressure of the gas in units such as atm, torr, mmHg, Pa
>
> V = volume of the gas in units such as L, mL
>
> n = number of moles of gas
>
> T = Kelvin temperature
>
> R = ideal gas constant: 0.0821 L·atm/mol·K (for most problems)

The value for R is 0.0821 L·atm/mol·K if the volume is expressed in liters, the pressure in atmospheres, and the temperature in Kelvin (naturally). You could calculate another ideal gas constant based on different units of pressure and volume, but the simplest thing to do is to use the 0.0821 and convert the given volume to liters and the pressure to atm. (The temperature must be in Kelvin in any case.)

EXAMPLE

▸ Let's see how we might use the ideal gas equation. Suppose you want to know what volume 40.0 g of hydrogen gas would occupy at 37.0°C and 0.850 atm? You have the pressure in atm, you can get the temperature in Kelvin (37.0°C + 273 = 310.0 K), but you will need to convert the grams of hydrogen gas to moles of hydrogen gas before you can use the ideal gas equation.

▸ First, you'll convert the 40.0 g to moles:

$$(40.0 \text{ g}) \times (1 \text{ mol } H_2/2.016 \text{ g}) = 19.84 \text{ mol } H_2$$

▸ (We're not worried about significant figures at this point since this is an intermediate calculation.)

EASY MISTAKE
Be sure to use the correct molecular mass for those gases that exist as diatomic molecules—H_2, N_2, O_2, F_2, Cl_2, Br_2, I_2.

▸ Now you can rearrange the ideal gas equation for the unknown quantity, the volume:

$$PV = nRT$$

$$V = nRT/P$$

▸ Finally, we enter in the numerical values for the different known quantities:

$$V = (19.84 \text{ mol}) (0.0821 \text{ L} \cdot \text{atm/mol} \cdot \text{K}) (310.0 \text{ K})/(0.850 \text{ atm})$$

$$V = 594 \text{ L (rounded to correct significant figures)}$$

Gas Stoichiometry

We can use the gas law relationships, especially the ideal gas law and the combined gas law, in reaction stoichiometry problems.

▸ For example, suppose you have 2.50 g of an impure sample of $KClO_3$ and you want to determine how many grams of pure $KClO_3$ are present. You heat the mixture and the $KClO_3$ decomposes according to the equation:

$$2 \text{ KClO}_3(s) \rightarrow 2 \text{ KCl}(s) + 3 \text{ O}_2(g)$$

▸ The oxygen gas that forms occupies a volume of 550.0 mL at 27°C. The atmospheric pressure is 731.3 torr.

▸ At this point, you now have 550.0 mL of oxygen gas at 731.3 torr and 300 K (27°C + 273). From this data, you can use the ideal gas equation to calculate the number of moles of oxygen gas produced:

$$PV = nRT$$

$$n = PV/RT$$

▶ You will need to convert the pressure from torr to atm:

$$(731.3 \text{ torr}) \times (1 \text{ atm}/760.0 \text{ torr})$$
$$= 0.9622 \text{ atm}$$

EASY MISTAKE

Be sure, especially in stoichiometry problems involving gases, that you are calculating the values such as volume and pressure of the correct gas. You can avoid this mistake by clearly labeling your quantities—that means, **mole of O_2** instead of just **mole**.

▶ and express the volume in liters:

$$550.0 \text{ mL} (1 \text{ L}/1000 \text{ mL}) = 0.5500 \text{ L}$$

▶ Now you can substitute these quantities into the ideal gas equation:

$$n = (0.9622 \text{ atm}) (0.5500 \text{ L}) / (0.0821 \text{ L} \cdot \text{atm}/\text{K} \cdot \text{mol}) (300 \text{ K})$$

$$n = 0.021486 \text{ moles } O_2 \text{ (unrounded)}$$

▶ Now you can use the reaction stoichiometry to convert from moles O_2 to moles $KClO_3$ and then to grams $KClO_3$:

$$\left(0.021486 \text{ mol } O_2\right)\left(\frac{2 \text{ mol } KClO_3}{3 \text{ mol } O_2}\right)\left(\frac{122.55 \text{ g } KClO_3}{1 \text{ mol } KClO_3}\right) = 1.7554$$
$$= 1.76 \text{ g } KClO_3$$

BTW

Check to make sure that your answer is reasonable. In this case, for example, the mass of the $KClO_3$ must be less than the mass of the impure mixture.

Another useful relationship is one derived from Avogadro's law: 1 mole of any gas occupies 22.4 L at STP (standard temperature and pressure of 0°C [273 K] and 1 atm). If you can find the volume at STP, you can then convert it to moles using this relationship and then to grams, if needed. However, if the value 22.4 L/mol is used, make sure that it is applied to a *gas* at STP.

BTW

Be sure when using any of the gas laws that you are dealing with gases, not liquids or solids.

The Kinetic-Molecular Theory of Gases

The **kinetic-molecular theory** (**KMT**) represents the properties of gases by modeling the gas particles themselves at the microscopic level. The KMT assumes that:

- Gases are composed of very small particles, either small molecules or individual atoms.

- The gas particles are so tiny in comparison to the distances between them that the KMT assumes the volume of the gas particles themselves is negligible.

- The gas particles are in constant motion, moving in straight lines in a random fashion and colliding with each other and the inside walls of the container. These collisions with the inside container walls comprise the pressure of the gas.

- KMT assumes that the gas particles neither attract nor repel each other. They may collide with each other, but if they do, it assumes the collisions are elastic—no kinetic energy is lost, only transferred from one gas molecule to another.

- Finally, the KMT assumes that the *average* kinetic energy of the gas is proportional to the Kelvin temperature.

A gas that obeys all these conditions is an **ideal gas**.

Using the KMT, we can derive several quantities related to the properties of the gas particles. First, KMT qualitatively describes the motion of the gas particles. The average velocity of the gas particles is the **root mean square velocity** and has the symbol u_{rms}. This is a special type of average speed. It is the average speed of a gas particle having the average kinetic energy. We can represent it as:

$$u_{rms} = \sqrt{3\ RT\ /\ M}$$

where R is a molar (ideal) gas constant (in terms of energy [joules] instead of liters and atmospheres) of 8.314 J/mol·K (= 8.314 kg·m^2/s^2 mol·K), T is the **Kelvin** temperature, and M is the molar mass of the gas. These root mean speeds are very high. Hydrogen gas, H_2, at 20°C has a value of approximately 2000 m/s.

Second, the KMT relates the average kinetic energy of the gas particles to the **Kelvin** temperature. We can represent the average kinetic energy per molecule as:

$$KE \text{ per molecule} = 1/2\ mv^2$$

where m is the mass of the molecule and v is its velocity.

We can represent the average kinetic energy per mole of gas as:

$$KE \text{ per mole} = 3/2\ RT$$

where R again is an ideal gas constant (8.314 J/mol·K) and T is the **Kelvin** temperature.

Graham's law defines the relationship of the speed of gas diffusion (mixing of gases due to their kinetic energy) or effusion (movement of a gas through a tiny opening) and the molecular mass. In general, the lighter the gas, the faster is its rate of effusion. Normally we use a comparison of the effusion rates of two gases with the specific relationship being:

$$r_1\ /\ r_2 = \sqrt{M_2\ /\ M_1}$$

where r_1 and r_2 are the rates of effusion/diffusion of gases 1 and 2, respectively, and M_2 and M_1 are the molecular (molar) masses of gases 2 and 1, respectively.

We can use Graham's law to determine the rate of effusion of an unknown gas knowing the rate of a known one, or we can use it to determine the molecular mass of an unknown gas.

For example, suppose you wanted to find the molar mass of an unknown gas. You measure its rate of effusion versus a known gas, H_2. The rate of hydrogen effusion was 3.728 mL/s, while the rate of the unknown gas was 1.000 mL/s. The molar mass of H_2 is 2.016 g/mol. Substituting into the Graham's law equation gives:

$$\frac{r_{H_2}}{r_{unk}} = \sqrt{\frac{M_{unk}}{2.016 \ \frac{g}{mol}}}$$

$$\frac{3.728 \ mL \ / \ s}{1.000 \ mL \ / \ s} = \sqrt{\frac{M_{unk}}{2.016 \ \frac{g}{mol}}}$$

$$M_{unk} = 28.018336 = 28.02 \ g/mol$$

BTW

Your answer must be reasonable. Hydrogen is the lightest gas (2.016 g/mol), so any molar mass less than this value is not reasonable.

Nonideal Gases

The KMT represents the properties of an ideal gas. However, there are no truly ideal gases; there are only gases that approach ideal behavior. We know that real gas particles do occupy a certain finite volume, and we know that there are interactions (attractions and repulsions) between real gas particles. These factors cause real gases to deviate a little from ideal behavior. However, a nonpolar gas at a low pressure and high temperature would come very close to ideal behavior. It would be nice, however, to have a more accurate model/equation for those times when we are doing extremely precise work, or we have a gas that exhibits a relatively large attractive or repulsive force. Johannes van der Waals introduced a modification of the ideal gas equation that attempted to consider the volume and attractive forces of real gases by introducing two constants a and b into the ideal gas equation. The result is the **van der Waals equation**:

$$(P + a\,n^2/V^2)\,(V - n\,b) = nRT$$

The attraction of the gas particles for each other tends to lessen the pressure of the gas, since the attraction slightly reduces the force of the collisions of the gas particles with the container walls. The amount of attraction depends on the concentration of gas particles and the magnitude of the intermolecular force of the particles. The greater the intermolecular forces of the gas, the higher the attraction is, and the less the real pressure. Van der Waals compensated for the attractive force by the term $(P + a\,n^2/V^2)$, where a is a constant for individual gases. The greater the attractive force between the molecules, the larger the value of a.

The actual volume of the gas is less than the ideal gas. This is because gas molecules do have a finite volume and the more moles of gas present, the smaller the real volume. The volume of the gas can be corrected by the $(V - n\,b)$ term, where n is the number of moles of gas and b is a different constant for each gas. The larger the gas particle, the more volume it takes up and the larger the b value.

The larger the gas particle, the more concentrated, and the stronger the intermolecular forces of the gas, the more deviation from the ideal gas equation one can expect and the more useful the van der Waals equation becomes.

Working Gas Law Problems

Gas law problems, like all problems, begin with isolating the variables and the unknown from the question. The usual suspects in gas law problems are pressure, volume, temperature, and moles. You will need to deal with at least two of these properties in every problem.

We will begin with a two-variable problem. A sample of a gas has a volume of 5.00 L at 25°C. What temperature, in °C, is necessary to increase the volume of the gas to 7.50 L?

We start by separating the numbers (and associated units) and the actual question from the remainder of the problem. Be very careful that the variables starting together stay together. Each column in the following table contains variables that began together.

V = 5.00 L	V = 7.50 L
T = 25°C = 298 K	T = ? °C

The presence of two volumes (or two temperatures) is a very strong indication that we will need to use the combined gas law. To use this gas law, we need subscripts to differentiate the different volumes and temperatures. Label one volume V_1 and the other V_2. It does not matter which volume we label 1 or 2 if we label all associated variables with the same subscript.

V_1 = 5.00 L	V_2 = 7.50 L
T_1 = 25°C = 298 K	T_2 = ? °C

The combined gas law is $(P_1V_1/T_1) = (P_2V_2/T_2)$. It is possible to simplify this equation in this problem by removing all variables not appearing in the table. The simplified combined gas law is $(V_1/T_1) = (V_2/T_2)$, which is a form of Charles's law. After simplification, we need to isolate the variable we are seeking (the one with the question mark in the table). Isolation of T_2 requires manipulating the equation. There are various

BTW

If you have a gas at a certain set of volume/temperature/pressure conditions and at least some of the conditions change, then you will probably be using the combined gas equation. If moles of gas are involved, the ideal gas equation will probably be necessary.

EASY MISTAKE

Change any other temperature units to Kelvin as soon as possible to minimize forgetting to do so later.

ways of doing this, all yielding the equation $T_2 = (T_1V_2/V_1)$. We now enter the appropriate values from our table into this equation:

$$T_2 = \frac{(T_1)(V_2)}{(V_1)} = \frac{(298 \text{ K})(7.50 \text{ L})}{(5.00 \text{ L})} = 447 \text{ K}$$

▶ To finish the problem, we must convert this Kelvin temperature to the requested Celsius temperature $T_2 = (447 - 273)°C = 174°C$.

BTW

The answer must make sense. In this case, there was an increase in volume, and, according to Charles's law, the temperature must also increase.

The next example shows what to do when the combined gas law cannot be simplified.

EXAMPLE

▶ Let's try another example: A sample of gas has a volume of 2.50 L at 25°C and 745 torr. What will be the volume, in liters, of the sample if we increase the temperature to 45°C and the pressure changes to 0.750 atm?

▶ As usual, we begin by extracting the numbers (and associated units) along with the desired unknown from the remainder of the problem.

$V_1 = 2.50 \text{ L}$ $\qquad$ $V_2 = ? \text{ L}$

$T_1 = 25°C = 298 \text{ K}$ $\qquad$ $T_2 = 45°C = 318 \text{ K}$

$P_1 = 745 \text{ torr}$ $\qquad$ $P_2 = 0.750 \text{ atm}$

▶ Note that the mandatory conversion to Kelvin is present. We will need to do a pressure conversion since P_1 and P_2 do not have the same units. We can do the pressure conversion at any time; however, we will temporarily postpone this step. The presence of two values for any variable strongly indicates that we need to use the combined gas law $[(P_1 V_1/ T_1) = (P_2 V_2/ T_2)]$. In this case, we cannot eliminate any of the variables. We will, therefore, go straight to the rearranging step to isolate V_2. This rearrangement gives $V_2 = (P_1 V_1 T_2 / T_1 P_2)$. We can now enter the values from the table into this equation. In addition, we will add a step to cover the pressure conversion.

$$V_2 = \frac{(P_1)(V_1)(T_2)}{(T_1)(P_2)} = \frac{(745 \text{ torr})(2.50 \text{ L})(318 \text{ K})}{(298 \text{ K})(0.750 \text{ atm})}\left(\frac{1 \text{ atm}}{760 \text{ torr}}\right)$$

$$= 3.4868 = 3.49 \text{ L}$$

IRL Many people get into trouble because they place conversions randomly about the page. If you do the conversion step as we did in this example, you will be less likely to "lose" your conversion, and less likely to copy a value incorrectly.

Here is an example in which we can use the ideal gas equation.

EXAMPLE

▶ Let's try another problem: What volume does 2.05 mol of oxygen occupy at a temperature of 25°C and a pressure of 0.950 atm?

▶ As usual, we begin by extracting the numbers (and associated units) and the desired unknown from the remainder of the problem:

$P = 0.950$ atm

$T = 25°C = 298$ K

$n = 2.05$ moles

$V = ?$

▶ Since a second set of conditions is not present, we will most likely need to use the ideal gas equation ($PV = nRT$). We need to rearrange this equation to isolate the unknown (V). Then we enter the appropriate values, including R, the ideal gas constant:

$$V = \frac{nRT}{P} = \frac{(2.05 \text{ mol})\left(0.0821 \dfrac{L \cdot atm}{mol \cdot K}\right)(298 \text{ K})}{(0.950 \text{ atm})}$$

$$= 52.7946 = 52.8 \text{ L}$$

Another type of gas law problem involves stoichiometry. Gas stoichiometry problems are just like all other stoichiometry problems—you must use moles. In addition, one or more gas laws are necessary.

> **EASY MISTAKE**
> For some reason, many students rearrange the ideal gas equation incorrectly; you should be very careful to check your rearrangement of this equation.

EXAMPLE

▶ Let's look at a gas stoichiometry problem. What volume, in liters of oxygen gas, collected over water, forms when 12.2 g of $KClO_3$ decompose according to the following equation:

$$2 \text{ } KClO_3(s) \rightarrow 2 \text{ } KCl(s) + 3 \text{ } O_2(g)$$

▶ The temperature of the water (and the gas) is 20.0°C, and the total pressure is 755 mmHg. (The vapor pressure of water at 20°C is given as 18 mmHg.)

▶ In this case, we not only need to separate the numbers (and associated units) but also, we need the balanced chemical equation.

$$2 \text{ } KClO_3(s) \qquad \rightarrow \qquad 2 \text{ } KCl(s) + 3 \text{ } O_2(g)$$

12.2 g   20.0°C = 293 K = T ? L = V

$$P_{total} = 755 \text{ mmHg}$$
$$P_{H_2O} = 18 \text{ mmHg}$$

▶ The presence of only one temperature, pressure, and volume indicates that we need to use the ideal gas law. To find the volume of *oxygen*, we need the pressure of *oxygen*, not the total pressure nor the vapor pressure of water. To find the oxygen pressure, we need a second gas law, Dalton's law. In this case, Dalton's law has the form $P_{total} = P_{O_2} + P_{H_2O}$, giving $P_{O_2} = P_{total} - P_{H_2O} = (755 - 18)$ mmHg = 737 mmHg.

▸ We can now rearrange the ideal gas law ($PV = nRT$) to isolate volume. Once rearranged, we can enter the values given:

$$V = \frac{nRT}{P} = \frac{(n)\left(0.0821\dfrac{L \cdot atm}{mol \cdot K}\right)(293\ K)}{(737\ mmHg)} =$$

▸ If it was not clear before, it should be clear now that we still must find moles. We will find moles from the mass of $KClO_3$ and the balanced chemical equation. We need to determine the molar mass of $KClO_3$ from the atomic weights of the individual elements (122.55 g/mol). We now add our mole information to the equation:

$$V = \frac{\left[(12.2\ g\ KClO_3)\left(\dfrac{1\ mol\ KClO_3}{122.55\ g\ KClO_3}\right)\left(\dfrac{3\ mol\ O_2}{2\ mol\ KClO_3}\right)\right]\left(0.0821\dfrac{L \cdot atm}{mol \cdot K}\right)(293K)}{(737\ mmHg)}$$

▸ To complete the problem, we need to add a pressure conversion:

$$V = \frac{\left[(12.2\ g\ KClO_3)\left(\dfrac{1\ mol\ KClO_3}{122.55\ g\ KClO_3}\right)\left(\dfrac{3\ mol\ O_2}{2\ mol\ KClO_3}\right)\right]\left(0.0821\dfrac{L \cdot atm}{mol \cdot K}\right)(293\ K)}{(737\ mmHg)}\left(\dfrac{760\ mmHg}{1\ atm}\right)$$

$$V = 3.70420 = 3.70\ L$$

▸ Notice that we could not use the relationship 22.4 L/mol, because this problem is not at STP.

The methods shown in this section will apply equally well to nonideal (real) gases, with the van der Waals equation used in place of the ideal gas equation. However, real gases require the use of van der Waals constants from appropriate tables.

BTW

You should be very careful when working problems involving gases and one or more other phases. The gas laws can only give direct information about gases. Therefore, there is a mole ratio conversion (from the balanced chemical equation) in this example to convert from the solid ($KClO_3$) to the gas (O_2).

EXERCISES

EXERCISE 5-1

Answer the following questions.

1. Write the expression for Dalton's law.

2. Write the expression for the combined gas equation.

3. Write the expression for the ideal gas equation.

4. List the pressure units introduced in this chapter and show the relationships between these units.

5. Write the rearranged expression of the ideal gas equation where you are solving for moles.

6. Write the rearranged expression of the combined gas equation where you are solving for T_2.

7. Fill in the blank in each of the columns below with either I (increases), or D (decreases), or C (constant). The potential changes apply to a sample of gas:

	a.	b.	c.	d.	e.
volume	constant		constant	constant	increases
pressure	increases	constant	increases		
temperature	constant	decreases		decreases	constant
moles		constant	constant	constant	constant

8. A sample of a gas occupies 15.55 L at a temperature of 27°C. What temperature, in °C, is necessary to adjust the volume of the gas to 10.00 L?

9. The initial pressure on a sample of oxygen gas was 795.0 torr. At this pressure, the sample occupied 1250.0 mL at 0°C. What was the final pressure, in atm, of this sample of oxygen gas if the final volume was 1.000 L and the final temperature was 25°C?

10. A sample of xenon gas was collected in a 5.000 L container at a pressure of 225.0 mmHg and a temperature of 27.0°C. Later the pressure had changed to 1.000 atm, and the temperature was 0.0°C. What was the new volume, in liters, of the gas?

11. The molecular weight of an unknown gas was to be determined through an effusion experiment. The unknown gas effused at a rate of 0.1516 mL/s. Under the same conditions, a sample of oxygen effused at a rate of 0.3238 mL/s. Determine the molecular weight of the unknown gas.

12. The decomposition of NaN_3 will generate N_2 gas. An 8.25 L sample of gas was collected over water at 25°C and at a total pressure of 875 torr. How many grams of NaN_3 reacted? The vapor pressure of water, at 25°C, is 24 torr.

$$2\ NaN_3(s) \rightarrow 2\ Na(s) + 3\ N_2(g)$$

Flashcard App

Thermochemistry

MUST ⚡ KNOW

⚡ A change in energy always accompanies physical and chemical changes. The heat change in a chemical reaction is the study of thermochemistry.

⚡ Calorimetry is the technique used to measure heat changes, enthalpies, and the types of heat capacities found in thermochemistry calculations.

⚡ Hess's law enables us to calculate the enthalpy change for a specific reaction.

In this chapter, we're going to look at the energy changes—especially heat—that occur during both physical and especially chemical changes. You might need to review the unit conversion method in Chapter 1 and the sections in Chapter 3 on balancing chemical reactions and the mole concept if you are not comfortable with them already. Let's get going.

Energy and Reactions

Thermochemistry deals with changes in heat that take place during chemical reactions. Heat is an **extensive property**, that is, it depends on the amount of matter (or the quantity of reactants that undergo change). Many times, we will be measuring the temperature (average kinetic energy) of the system. Temperature is an **intensive property**, one that is independent of the amount of matter present. We will be discussing the energy exchanges between the system that we are studying and the surroundings. The **system** is that part of the universe that we are studying. It may be a beaker, or it may be the solar system. The **surroundings** are the rest of the universe that the change affects.

The most common units of energy that we use in the study of thermodynamics are the joule and the calorie. The **joule (J)** is:

$$1 \text{ J} = 1 \text{ kg} \cdot \text{m}^2/\text{s}^2$$

The **calorie (cal)** is the amount of energy needed to raise the temperature of 1 g of water by 1°C and relates to the joule as:

$$1 \text{ cal} = 4.184 \text{ J}$$

 IRL This is not the same calorie that is commonly associated with food and diets. That is the nutritional Calorie (Cal), which is really a kilocalorie (1 Cal = 1000 cal). That 300 Cal candy bar really contains 300,000 cal of energy.

Calorimetry is the laboratory technique used to measure the heat released or absorbed during a chemical or physical change. The quantity of heat absorbed or released during the chemical or physical change is q and is proportional to the change in temperature of the system. This system has a **heat capacity**, which is the quantity of heat needed to change the temperature 1 K. It has the form:

heat capacity = $q/\Delta T$

The heat capacity most commonly has units of J/K. The **specific heat capacity (or specific heat) (c)**, is the quantity of heat needed to raise the temperature of 1 g of the substance 1 K:

$c = q/(\text{mass})(\Delta T)$ or $q = (c)(\text{mass})(\Delta T)$

The specific heat capacity commonly has units of J/g·K. The specific heat capacity of water is 4.18 J/g·K = 4.18 J/g·°C. If we have the specific heat capacity, the mass, and the change of temperature, it is possible to determine the amount of energy absorbed or released (q).

BTW

The specific heat capacity of water is necessary to solve many problems. However, when reading the problem, it is often not apparent that you must use this value. If the problem mentions water, its specific heat capacity will often be part of the solution to the problem.

Another related quantity is the **molar heat capacity (C)**, the amount of heat needed to change the temperature of 1 mol of a substance 1 K.

Calorimetry involves the use of a laboratory instrument called a **calorimeter**. Two types of calorimeters are commonly used, a simple coffee-cup calorimeter and a more sophisticated bomb calorimeter. In both, we carry out a reaction with known amounts of reactants and the change in temperature is measured.

BTW

Don't mix energy units, J and cal; always use appropriate conversions.

This is a coffee-cup calorimeter:

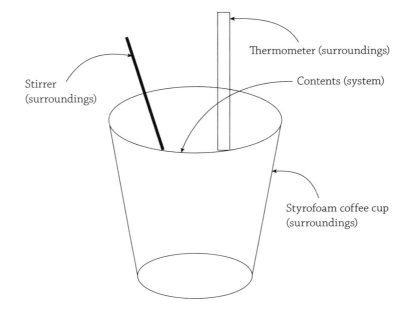

Thermometer (surroundings)

Stirrer
(surroundings)

Contents (system)

Styrofoam coffee cup
(surroundings)

The coffee-cup calorimeter can be used to measure the heat changes in reactions that are open to the atmosphere, q_p, constant pressure reactions. We use this type of calorimeter to measure the specific heats of solids. We heat a known mass of a substance to a certain temperature and then add it to the calorimeter containing a known mass of water at a known temperature. The final temperature is then measured. We know that the heat lost by the added substance (the system) is equal to the heat gained by the surroundings (the water and calorimeter, although for simple coffee-cup calorimetry the heat gained by the calorimeter is small and often ignored):

$$-q_{solid} = q_{water}$$

Substituting the relationships for q gives:

$$- (c_{solid} \times mass_{solid} \times \Delta T_{solid}) = (c_{water} \times mass_{water} \times \Delta T_{water})$$

We can then solve this equation for the specific heat capacity of the solid.

We use the constant-volume bomb calorimeter to measure the energy changes that occur during combustion reactions. We add a weighed sample of the substance under investigation to the calorimeter and then add excess compressed oxygen gas. We ignite the sample electrically and measure the temperature change of the calorimeter and the known mass of water. Generally, we know the heat capacity of the calorimeter and can determine heat absorbed by the water. Suppose we have a problem such as the following.

EXAMPLE

▶ We ignited a 1.5886 g sample of glucose ($C_6H_{12}O_6$) in a bomb calorimeter. The temperature increased by 3.682°C. The heat capacity of the calorimeter was 3.562 kJ/°C, and the calorimeter contained 1.000 kg of water. Find the molar heat of reaction (J/mol of glucose) for the reaction:

$$C_6H_{12}O_6(s) + 6\ O_2(g) \rightarrow 6\ CO_2(g) + 6\ H_2O(l)$$

▶ In solving problems of this type, you must realize that the oxidation of the glucose released energy in the form of heat and that some of the heat was absorbed by the water and the remainder by the calorimeter. You can use both the heat capacity of the calorimeter and the mass and specific heat of the water with the temperature change to calculate the heat absorbed by the calorimeter and water:

heat absorbed by the calorimeter: (3.562 kJ/°C) × 3.682°C

= 13.12 kJ

heat absorbed by the water: 1.000 kg × (1000 g/kg) ×
(4.184 J/g°C) × (1 kJ/1000 J) × 3.682°C = 15.40 kJ

total heat absorbed by the calorimeter and water =
total heat released by the glucose

↓

13.12 kJ + 15.40 kJ = 28.52 kJ

▸ Converting the grams of glucose to moles then allows us to calculate the molar heat capacity:

Heat released by the glucose/moles glucose =
molar heat of reaction

↓

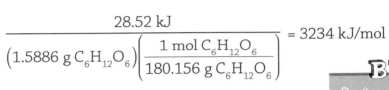

$$\frac{28.52 \text{ kJ}}{\left(1.5886 \text{ g } C_6H_{12}O_6\right)\left(\dfrac{1 \text{ mol } C_6H_{12}O_6}{180.156 \text{ g } C_6H_{12}O_6}\right)} = 3234 \text{ kJ/mol}$$

▸ You will need to change this value to a negative value because the reaction gave off this quantity of heat. Thus the heat of reaction is: −3234 kJ/mol $C_6H_{12}O_6$.

BTW

Don't confuse the heat capacity, specific heat, and the molar heat capacity (watch the units).

Enthalpy (ΔH)

Many of the reactions that chemists study are reactions that occur at constant pressure. Because this constant pressure situation is so common in chemistry, scientists use a special thermodynamic term to describe this energy, **enthalpy, H**. The **enthalpy change, ΔH**, is equal to the heat gained or lost by the system under constant pressure conditions. The following sign conventions apply:

If $\Delta H > 0$ the reaction is endothermic

If $\Delta H < 0$ the reaction is exothermic

EASY MISTAKE ΔH is dependent on the state of matter. The enthalpy change would be different for the formation of liquid water instead of gaseous water.

We sometimes call it the $\Delta H_{reaction}$ (ΔH_{rxn}). The ΔH is normally associated with a specific reaction. For example, the enthalpy change associated with the formation of hydrogen and oxygen gases from water vapor is:

$$2 \, H_2O(g) \rightarrow 2 \, H_2(g) + O_2(g) \quad \Delta H = +483.6 \text{ kJ}$$

The positive sign indicates that this reaction is endothermic. This value of ΔH is for the decomposition of 2 moles of water. If 4 moles were decomposed, the ΔH would be 2×483.6 kJ.

BTW *Watch your signs in all the thermodynamic calculations. They are extremely important.*

If we reverse the reaction for the decomposition of water above, the sign of the ΔH reverses, -483.6 kJ. That would indicate that the reaction releases 483.6 kJ of energy in forming 2 moles of water. This would now become an exothermic process.

Hess's Law

We can measure enthalpies of reaction using a calorimeter. However, we can also calculate the values. **Hess's law** states that if we express a reaction in a series of steps, then the enthalpy change for the overall reaction is simply the sum of the enthalpy changes of the individual steps. If, in adding the equations of the steps together, it is necessary to reverse one of the given reactions, then we will need to reverse the sign of the ΔH. In addition, we must pay attention if we must adjust the reaction stoichiometry.

It really doesn't matter whether the steps which are used are the actual ones in the mechanism (pathway) of the reaction because $\Delta H_{reaction}$ is a **state function**, a function that doesn't depend on the pathway, only the initial and final states.

▶ Given the following information:

$$C(s) + O_2(g) \rightarrow CO_2(g) \qquad \Delta H = -393.5 \text{ kJ}$$

$$H_2(g) + (1/2) O_2(g) \rightarrow H_2O(l) \qquad \Delta H = -285.8 \text{ kJ}$$

$$C_2H_2(g) + (5/2) O_2(g) \rightarrow 2 CO_2(g) + H_2O(l) \qquad \Delta H = -1299.8 \text{ kJ}$$

▶ Find the enthalpy change for: $2 C(s) + H_2(g) \rightarrow C_2H_2(g)$.

▶ Since we need 2 carbon atoms, we will multiply the first equation by 2:

$$2 (C(s) + O_2(g) \rightarrow CO_2(g)) \qquad 2(-393.5 \text{ kJ})$$

$$H_2(g) + (1/2) O_2(g) \rightarrow H_2O(l) \qquad -285.8 \text{ kJ}$$

▶ Since C_2H_2 appears on the product side, we will reverse the third reaction and change the sign of ΔH:

$$2CO_2(g) + H_2O(l) \rightarrow C_2H_2(g) + (5/2) O_2(g) \quad -(-1299.8 \text{ kJ})$$

▶ The equations are now:

$$2 C(s) + 2 O_2(g) \rightarrow 2 CO_2(g) \qquad -787.0 \text{ kJ}$$

$$H_2(g) + (1/2) O_2(g) \rightarrow H_2O(l) \qquad -285.8 \text{ kJ}$$

$$2 CO_2(g) + H_2O(l) \rightarrow C_2H_2(g) + (5/2) O_2(g) \quad +1299.8 \text{ kJ}$$

▶ Adding the reactions and ΔHs and cancelling any species appearing on both sides of the reaction arrows:

$$2 C(s) + H_2(g) \rightarrow C_2H_2(g) \qquad 227.0 \text{ kJ}$$

▶ Note that there was $(2 + 1/2) O_2(g)$ on the left side and $(5/2) O_2(g)$ on the right side; these are equal, and they cancel. Other things that cancel are $2 CO_2(g)$ and $H_2O(l)$, because there are equal quantities on the reactant and product side. Do not forget, only identical species can cancel.

Calorimetry

Calorimetry problems appear in the thermochemistry chapter of many texts. These problems often appear intimidating, but this need not be the case.

 IRL   Many people have trouble with the problems concerning this material. The major cause is that many of the problems are too simple. You need to analyze carefully what the questions are really asking. Unit conversions are exceedingly important in working these problems.

EXTRA HELP   A+

EXAMPLE

> A sample of $C_2H_5OH(l)$ weighing 1.42 g was burned with excess $O_2(g)$ in a bomb calorimeter. After completion of the reaction, the temperature of the calorimeter had increased from 24.00°C to 28.36°C. The calorimeter contained 0.500 kg of water. The heat capacity of the calorimeter was 7.54 kJ/°C. Calculate the heat of reaction in kilojoules per mole of C_2H_5OH. The reaction was:

$$C_2H_5OH(l) + 3\ O_2(g) \rightarrow 2\ CO_2(g) + 3\ H_2O(l)$$

> The question is "Find the molar heat of reaction." This means we need the enthalpy change (ΔH) in kJ/mol (or J/mol) of C_2H_5OH. The increase in temperature means that this is an exothermic process (negative enthalpy change).

> The presence of water implies that we may need some properties of water. In this case, we will need the specific heat of water (4.18 J/g°C).

BTW

The presence of water in any problem may require various unspecified properties of water. The most common unspecified values are the specific heat of water, as in this problem, and the density of water.

▶ Extracting the information from the problem gives:

$$C_2H_5OH(l) + 3\ O_2(g) \rightarrow 2\ CO_2(g) + 3\ H_2O(l)$$

1.42 g   24.00°C → 28.36°C

7.54 kJ/°C

0.500 kg H_2O

4.18 J/g°C

▶ We need the temperature change, not the separate temperatures. The temperature change is:

$$\Delta T = T_{final} - T_{initial} = 28.36°C - 24.00°C = 4.36°C$$

▶ We will begin by finding the quantity of energy involved in the reaction. The energy produced did two things: part of the energy warmed the calorimeter, and the remainder of the energy warmed the water. We need to determine these two values separately and then combine them to determine the total energy change. The amount of energy absorbed by the calorimeter is:

$$\Delta H_{calorimeter} = \left(\frac{7.54\ kJ}{°C}\right)(4.36\,°C) = 32.8744\ kJ\ (\text{unrounded})$$

▶ The amount of energy absorbed by the water is:

$$\Delta H_{water} = \left(\frac{4.18\ J}{g\,°C}\right)(0.500\ kg)(4.36°C)\left(\frac{1000\ g}{1\ kg}\right)$$

$$= 9112.4\ J\ (\text{unrounded})$$

▶ The total amount of heat energy will be the sum of these values ($\Delta H_{calorimeter} + \Delta H_{water}$). However, before we can sum these values, we must make sure the units match. We can either convert the kilojoules to joules or the joules to kilojoules. In this case we will convert the joules to kilojoules and then add the values:

$$\Delta H_{total} = \Delta H_{calorimeter} + \Delta H_{water} = 32.8744 \text{ kJ}$$

$$+ 9112.4 \text{ J} \left(\frac{1 \text{ kJ}}{10^3 \text{ J}} \right) = 41.9868 \text{ kJ}$$

▸ We have already noted that the process was exothermic (the value is negative) and should be reported as –41.9868 kJ (unrounded). The sign comes from convention (the definition of exothermic) and not from the calculation.

▸ To get the final answer, we must divide the energy released by the moles of C_2H_5OH. We find the moles of C_2H_5OH from the mass given in the problem and the molar mass determined from the atomic weights of the elements:

$$(1.42 \text{ g } C_2H_5OH) \left(\frac{1 \text{ mol } C_2H_5OH}{46.07 \text{ g } C_2H_5OH} \right) = 3.082333 \times 10^{-2} \text{ mole}$$

$$C_2H_5OH \text{ (unrounded)}$$

▸ Combining the enthalpy change with the moles gives us the final answer:

$$\Delta H_{reaction} = \left(\frac{-41.9868 \text{ kJ}}{3.082333 \times 10^{-2} \text{ mol}} \right) = -1362.18$$

$$= -1.36 \times 10^3 \text{ kJ/mol}$$

BTW

Some problems use the total heat capacity of the calorimeter (calorimeter + water). This means that the summation has already been done for you, so you do not need to do the summation a second time.

Calorimetry problems, such as this one and most other thermochemistry problems, will require a systematic step-by-step approach as seen in this example.

EXERCISES

EXERCISE 6-1

Answer these questions about thermochemistry.

1. True or False: Temperature is an extensive property.

2. True or False: A nutritional Calorie is the same as a "normal" calorie.

3. True or False: One way to determine the heat of reaction is to use a calorimeter.

4. True or False: An exothermic process has a negative enthalpy.

5. True or False: Water cooling from 40°C to 25°C is an exothermic process.

6. Write the definition of a joule in terms of SI base units.

7. What are the units of specific heat capacity?

8. Calculate the standard heat of formation for $HC_2H_3O_2(l)$ using the following:

$$C(s) + O_2(g) \rightarrow CO_2(g) \qquad \Delta H = -393.5 \text{ kJ}$$

$$H_2(g) + 1/2 \, O_2(g) \rightarrow H_2O(l) \qquad \Delta H = -285.8 \text{ kJ}$$

$$HC_2H_3O_2(l) + 2 \, O_2(g) \rightarrow 2 \, CO_2(g) + 2 \, H_2O(l) \qquad \Delta H = -871 \text{ kJ}$$

The reaction for the standard heat of formation of $HC_2H_3O_2(l)$ is:

$$2 \, C(s) + 2 \, H_2(g) + O_2(g) \rightarrow HC_2H_3O_2(l)$$

9. Calculate the heat of reaction for $2 N_2(g) + 5 O_2(g) \rightarrow 2 N_2O_5(g)$ using the following thermochemical equations:

$$N_2(g) + 3 O_2(g) + H_2(g) \rightarrow 2 HNO_3(aq) \qquad \Delta H = -413.14 \text{ kJ}$$

$$N_2O_5(g) + H_2O(g) \rightarrow 2 HNO_3(aq) \qquad \Delta H = 218.4 \text{ kJ}$$

$$2 H_2(g) + O_2(g) \rightarrow 2 H_2O(g) \qquad \Delta H = -483.64 \text{ kJ}$$

10. Propane gas, C_3H_8, is sometimes used as a fuel. To measure its energy output as a fuel, a 1.860 g sample was combined with an excess of O_2 and ignited in a bomb calorimeter. After the reaction, it was found that the temperature of the calorimeter had increased from 25.000°C to 26.061°C. The calorimeter contained 1.000 kg of water. The heat capacity of the calorimeter was 4.643 kJ/°C. Determine the heat of reaction for the reaction in kJ/mole propane. The reaction was: $C_3H_8(l) + 5 O_2(g) \rightarrow 3 CO_2(g) + 4 H_2O(l)$.

Electrons and Quantum Theory

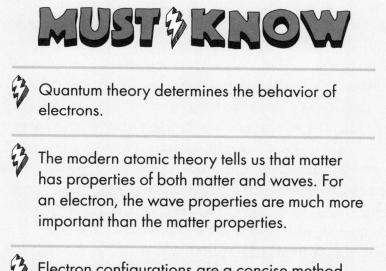

MUST KNOW

- Quantum theory determines the behavior of electrons.

- The modern atomic theory tells us that matter has properties of both matter and waves. For an electron, the wave properties are much more important than the matter properties.

- Electron configurations are a concise method of describing the arrangement of electrons in an atom.

ow we'll investigate electrons and the current models for where those electrons are located within the atom. You may want to briefly review Chapter 2 concerning electrons, protons, and neutrons.

Light and Matter

In the early development of the atomic model, scientists initially thought that they could define the subatomic particles by the laws of classical physics—that is, they were tiny bits of matter. However, they later discovered that this particle view of the atom could not explain many of the observations that scientists were making. About this time, a model (the quantum mechanical model) that attributed the properties of both matter and waves to particles began to gain favor. This model described the behavior of electrons in terms of waves (electromagnetic radiation).

Light, or radiant energy, makes up the **electromagnetic spectrum**. Light includes gamma rays, x-rays, ultraviolet, visible, and so on. The energy of the electromagnetic spectrum moves through space as waves that have three associated properties: frequency, wavelength, and amplitude. The **frequency**, ν, is the number of waves that pass a point per second. The **wavelength,** λ, is the distance between two identical points on a wave, while **amplitude (or peak amplitude)**, A, is the height of the wave.

Wave Diagram

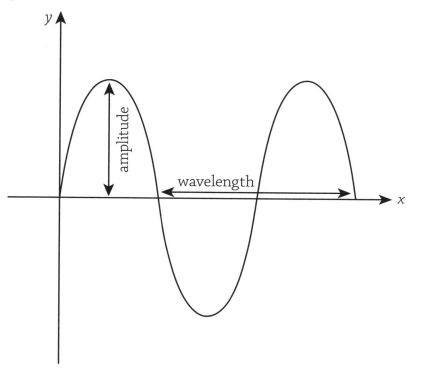

We define the energy associated with a certain frequency of light by the equation $E = h\nu$ where h is Planck's constant = 6.63×10^{-34} J • s.

In the development of the quantum mechanical model of the atom, scientists found that an electron in an atom could have only certain distinct quantities of energy associated with it and that to change its energy it had to absorb or emit a certain distinct amount of energy. The energy that the atom emits or absorbs is really the difference in the two energy states, and we can calculate it by the equation:

$$\Delta E = h\nu$$

Electromagnetic radiation travels at about the same speed in a vacuum, 3.00×10^8 m/s. This constant, c, is the **speed of light** and is the product of the frequency and the wavelength:

$$c = \lambda \nu$$

Therefore, we can represent the energy change in terms of the wavelength and speed of light:

$$\Delta E = h\, c / \lambda$$

Bohr's Model

Niels Bohr developed the first modern atomic model for hydrogen using the concepts of quantized energies—the energies associated with the atom could only be of certain discrete values. The Bohr model postulated a **ground state** for the atom, an energy state of lowest energy, and one or more **excited states**, energy states of higher energy. For an electron in an atom to go from its ground state to an excited state, it must absorb a certain amount of energy, and if the electron dropped back from that excited state to its ground state, it must emit that same amount of energy. Bohr's model also allowed the development of a method of calculating the energy difference between any two energy levels:

$$\Delta E = -2.18 \times 10^{-18} \text{ J} \left(\frac{1}{n_{final}^2} - \frac{1}{n_{initial}^2} \right)$$

The constant 2.18×10^{-18} J is the Rydberg constant, R_H. The ns are integers associated with the initial and final energy levels.

Quantum Mechanics

Bohr's model worked relatively well for hydrogen, but not very well at all for any other atom. Early in the 1900s, Erwin Schrödinger created a more detailed model and set of equations that better described atoms by using quantum mechanical concepts. His model introduced a mathematical description of the electron's motion called a **wave function** or **atomic orbital**. Squaring the wave function (orbital) gives the volume of space in which the probability of finding the electron is high, the **electron cloud** (**electron density**).

Schrödinger's equation required the use of **quantum numbers** to describe each electron within an atom corresponding to the orbital size, shape, and orientation in space. Later it was found that one needed a quantum number associated with the electron spin.

Louis de Broglie, in the mid-1920s, proposed the idea that particles could be treated as waves by the relationship

$$\lambda = h/mv$$

This equation related the mass (m) and velocity (v not ν) of a particle to its wavelength (λ) by using Planck's constant (h).

Quantum Numbers and Orbitals

The first quantum number is the **principal quantum number (n)** that describes the size and energy of the orbital and relative distance from the nucleus. The possible values of n are positive integers (1, 2, 3, 4, and so on). The smaller the value of n, the lower the energy, and the closer the orbital is to the nucleus. We sometimes refer to the principal quantum number as designating the **shell** the electron is occupying.

Each shell contains one or more subshells, each with one or more orbitals. The second quantum number is the **angular momentum quantum number (l)** that describes the shape of the orbitals. Its value is related to the principal quantum number and has allowed values of 0 to $(n - 1)$. For example, if $n = 4$, then the possible values of l would be 0, 1, 2, and 3 ($= 4 - 1$).

- If $l = 0$, then the orbital is called an s-orbital and has a spherical shape with the nucleus at the center of the sphere. The greater the value of n, the larger is the sphere.

- If $l = 1$, then the orbital is called a p-orbital with two lobes of high electron density on either side of the nucleus, for an hourglass or dumbbell shape.

- If $l = 2$, then the orbital is a d-orbital with a variety of shapes.

- If $l = 3$, then the orbital is an f-orbital with more complex shapes.

BTW

The procedure here locates the electrons in terms of energy, not position. The Heisenberg uncertainty principle says that it is impossible to determine both the position (location) and momentum of an electron.

The s-, p-, and d-Orbitals

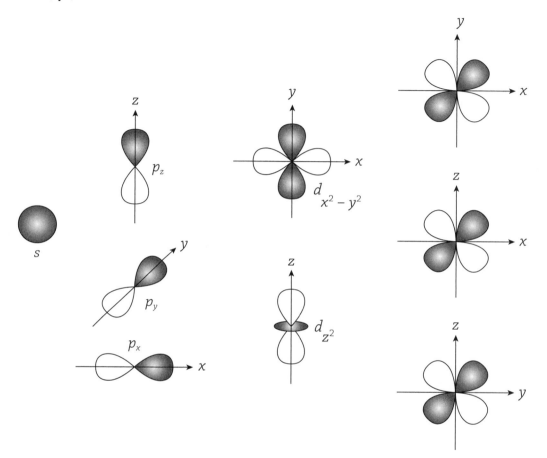

The third quantum number is the **magnetic quantum number** (m_l). It describes the orientation of the orbital around the nucleus. The possible values of m_l depend on the value of the l quantum number. The allowed values for m_l are $-l$ through 0 to $+l$. For example, for $l = 3$, the possible values of m_l would be $-3, -2, -1, 0, +1, +2, +3$. This is why, for example, if $l = 1$ (a p-orbital), there are three p-orbitals (sublevels) corresponding to m_l values of $-1, 0, +1$.

The fourth quantum number is the **spin quantum number** (m_s) and indicates the direction the electron is spinning. There are only two possible values for m_s: +1/2 and –1/2. When two electrons are to occupy the same orbital, then one must have an m_s = +1/2 and the other electron must have an m_s = –1/2. These are spin-paired electrons.

To assign the four quantum numbers for an electron, first begin with the electron in the lowest energy level, $n = 1$. Assign the value of n, then the corresponding values of l, m_l, and finally m_s. Once you have finished all the possible electrons at $n = 1$, repeat the procedure with $n = 2$. Don't forget about Hund's rule and the Pauli exclusion principle (both discussed later). The quantum numbers for the six electrons in carbon would be:

Quantum Number	First Electron	Second Electron	Third Electron	Fourth Electron	Fifth Electron	Sixth Electron
n	1	1	2	2	2	2
l	0	0	0	0	1	1
m_l	0	0	0	0	–1	0
m_s	+ ½	– ½	+ ½	– ½	+ ½	+ ½

Electron Configuration

Quantum mechanics may be used to determine the arrangement of the electrons within an atom if two specific principles are applied: the Pauli exclusion principle and the aufbau principle. The **Pauli exclusion principle** states that no two electrons in an atom can have the same set of the four quantum numbers. For example, if an electron has the following set of quantum numbers: $n = 1$, $l = 0$, $m_l = 0$, and m_s = +1/2, then no other electron in that atom may have the same set. The Pauli exclusion principle limits all orbitals to only two electrons. For example, the 1s-orbital is filled when it has two electrons, so that any additional electrons must enter another orbital.

The second principle, the **aufbau principle**, describes the order in which the electrons enter the different orbitals and sublevels. The arrangement of electrons builds up from the lowest energy level. The most stable arrangement of electrons has all the electrons with the lowest possible energy. This lowest energy arrangement is the **ground state**. Less stable (higher energy) arrangements are the **excited states**. An atom may have any number of excited state arrangements, but there is only one ground state.

When following the aufbau principle, the orbitals begin filling at the lowest energy and continue to fill until we account for all the electrons in an atom. Filling begins with the $n = 1$ level followed by the $n = 2$ level, and then the $n = 3$ level. However, there are exceptions in this sequence. In addition, **Hund's rule** states that the sublevels within an orbital will half fill before the electrons pair up in a sublevel.

The exceptions begin with the fourth energy level. The fourth energy level begins to fill before all the sublevels in the third shell are complete. More complications in the sequence appear as the value of the principle quantum number increases. The sequence of orbital filling, with complications, is: $1s$, $2s$, $2p$, $3s$, $3p$, $4s$, $3d$, $4p$, $5s$, $4d$, $5p$, $6s$, $4f$, $5d$, $6p$, $7s$, $5f$, $6d$, and so on.

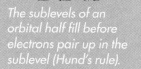

BTW

The sublevels of an orbital half fill before electrons pair up in the sublevel (Hund's rule).

The figure on the next page illustrates the aufbau principle diagrammatically. The orbitals begin filling from the bottom of the diagram (lowest energy) with two electrons maximum per individual sublevel (line on the diagram).

The atomic orbitals in order of increasing energy

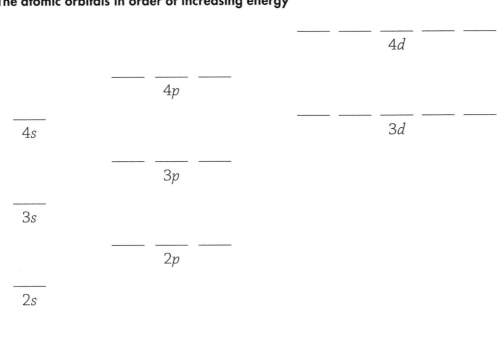

There are several ways of indicating the arrangement of the electrons in an atom. The most common way is the electron configuration. The **electron configuration** requires the use of the n and l quantum numbers along with the number of electrons. The principal quantum number, n, is represented by an integer (1, 2, 3 . . .), and a letter represents the l quantum number ($0 = s$, $1 = p$, $2 = d$, and $3 = f$). Any s-subshell can hold a maximum of 2 electrons, any p-subshell can hold up to 6 electrons, any d-subshell can hold a maximum of 10 electrons, and any f-subshell can hold up to 14 electrons.

EXAMPLE

▸ Lithium has three electrons. What would its electron configuration be?

▸ Having three electrons, lithium has two electrons in the 1s-orbital and one in the 2s orbital. Its electron configuration then would be $1s^2 2s^1$.

The electron configuration for fluorine (nine electrons) is: $1s^2 2s^2 2p^5$. The figure below shows one way of remembering the pattern for filling the atomic orbitals. The filling begins at the top of the pattern and follows the first arrow. When you reach the end of the first arrow, you go to the second arrow and follow it to the end. The third arrow continues the pattern, and so forth:

BTW

The sum of the superscripts must equal the total number of electrons present.

Filling Atomic Orbitals

BTW

Maximum number of electrons for s-subshells = 2, p-subshells = 6, d-subshells = 10, f-subshells = 14.

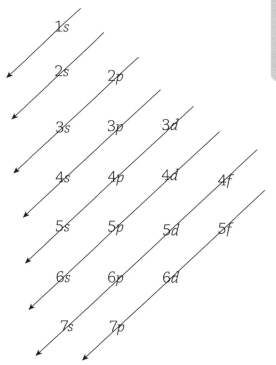

▶ Let's use cobalt (Co) to illustrate how to apply the aufbau principle. Cobalt has 27 electrons (refer to its atomic number on the periodic table, on page 444). Beginning at the top of the figure on the previous page, we see that:

Filling Atomic Orbitals

The first 2 electrons will fill the 1s sublevel to give $1s^2$, leaving 25 electrons.

$\downarrow$

The next 2 electrons will enter the 2s sublevel giving $1s^2 2s^2$, leaving 23 electrons.

$\downarrow$

Next is the 2p sublevel, which uses 6 electrons to give $1s^2 2s^2 2p^6$ and leaving 17 electrons.

$\downarrow$

After the 2p is the 3s sublevel, which takes 2 electrons, $1s^2 2s^2 2p^6 3s^2$, leaving 15 electrons.

$\downarrow$

Filling the 3p sublevel gives $1s^2 2s^2 2p^6 3s^2 3p^6$ with 9 electrons to go.

$\downarrow$

The 4s sublevel follows the 3p, giving $1s^2 2s^2 2p^6 3s^2 3p^6 4s^2$ and leaving 7 electrons.

$\downarrow$

Finally, we reach the 3d sublevel, which can hold 10 electrons. Since we only have seven electrons remaining, the 3d sublevel will not fill.

▶ The ground-state electron configuration for cobalt is, therefore, $1s^2 2s^2 2p^6 3s^2 3p^6 4s^2 3d^7$.

There are a few exceptions to this general pattern. The two best-known exceptions are the elements chromium, Cr, and copper, Cu. The electron configurations for these two elements are:

Cr $1s^22s^22p^63s^23p^64s^13d^5$

Cu $1s^22s^22p^63s^23p^64s^13d^{10}$

In both cases, the $4s$ sublevel is not filled even though there are electrons in the $3d$ sublevel. In a later chapter, we will see why these "exceptions" are predictable.

In some cases, it would be convenient to condense the electron configuration. In this condensed form, the electronic configuration of the *previous* noble gas forms a core represented by the atomic symbol of the element enclosed in brackets (i.e., [He] or [Ne]). The electrons added since the noble gas, follow the noble gas core. For example, cobalt can be represented as an argon core plus the $4s$ and $3d$ electrons. Thus, $1s^22s^22p^63s^23p^64s^23d^7$ becomes $[Ar]4s^23d^7$.

The electrons that are contained in the noble gas core are the **core electrons** while the electrons outside the core are **valence electrons**. These valence electrons are involved in the chemical behavior of the elements. For the representative elements, the valence electrons are those s and p electrons in the outermost energy level. The **valence shell** contains the valence electrons.

BTW

It is only the nearest **previous** *noble gas that is used as the core.*

Working Matter and Light Problems

If you are going to be calculating any values concerning matter, light, and energy, it will help to make a list of the equations. These equations are:

$$c = \lambda \nu$$

$$\lambda = h / mv$$

$$E = h\nu$$

$$E = hc/\nu \; (\Delta E \text{ may replace } E)$$

$$\Delta E = -2.18 \times 10^{-18} \text{ J} \left(\frac{1}{n^2_{\text{final}}} - \frac{1}{n^2_{\text{initial}}} \right)$$

These problems require two constants:

Planck's constant $\quad h = 6.63 \times 10^{-34}$ J·s

the speed of light $\quad c = 3.00 \times 10^{8}$ m/s

EASY MISTAKE
Many students get into trouble because they confuse velocity, v, with frequency, ν. Make sure you are not one of these students.

You may wish to memorize these constants; however, through repeated use you will probably learn these values without any further "memorization."

EASY MISTAKE
Some people mistakenly list the units of Planck's constant as J/s. You should be careful not to be one of these individuals.

When beginning a problem of this type you should, as always, carefully extract the values (including units) from the question. You should then label each of the extracted values with the appropriate symbol from the equation set. To help in the assignment, pay attention to the units. The energy, E, will have units of J or kJ. The frequency, ν, will have units of $1/s = s^{-1} = $ Hz. The wavelength, λ, will always have units of length, usually m or nm, but also English units or Å (angstrom). Mass, m, may have any mass unit, and you will need to change this mass to kilograms. Velocity, v, may have any units of distance (length) divided by any unit of time. In most cases, you will need to convert the velocity to m/s.

Once you have extracted and labeled the values in the problem (including the value being sought), you finish the problem by matching the symbols you have with the given equations. After rearranging the equation, if necessary, you will enter the appropriate values and conversions to get the final answer.

EXERCISES

EXERCISE 7-1

Answer the following questions.

1. Write the equation that relates energy to frequency.

2. How many different ground states may an atom have?

3. What are the maximum number of electrons that may occupy *s*-, *p*-, *d*-, and *f*-orbitals?

4. List the four quantum numbers and give their symbols.

5. State Hund's rule.

6. Give the full electron configuration of each of the following.
   a. F
   b. Cu
   c. Re

7. A laser emits light with a wavelength of 645 nm. What is the frequency of this light?

8. An *alpha* particle (mass = 6.6×10^{-24} g) emitted by radium travels at 5.5×10^{10} m/h. What is its de Broglie wavelength (in meters)?

9. It requires 239 kJ/mol to separate the chlorine atoms in a Cl_2 molecule. What wavelength of light would be necessary to separate the atoms in a single chlorine molecule?

Flashcard App

8 Periodic Trends

MUST KNOW

 The elements on the periodic table exhibit trends in their properties.

The ionization energy (IE) is the energy required to remove an electron from a gaseous atom in its ground state.

The energy change accompanying the addition of an electron to a gaseous atom in its ground state is the electron affinity (EA).

W e're going to look further into the relationship between the properties of elements and their position on the periodic table. These properties include ionization energies and electron affinities. You may want to review the basic structure of the periodic table in Chapter 2 and electron configurations in Chapter 7. Let's start with renewing our acquaintance with the periodic table.

Periodic Table Revisited

In Chapter 2, we showed you the arrangement of the periodic table. The columns are groups or families. The group members have similar chemical properties and somewhat similar physical properties. The rows are periods. The row members have predictably different properties. In Chapter 7, we showed you the electron configurations of the elements. You learned how similar electron configurations lead to similar chemical properties. There are, however, other periodic properties. These properties relate to the magnitude of the attractive force that the protons in the nucleus have for electrons. These include atomic radii, ionization energies, and electron affinities. The inner electrons (core electrons) somewhat screen the valence electrons from the attraction of the nucleus. The **effective nuclear charge** is the nuclear charge minus the screening effect of the core electrons. This screening is essentially constant for all the elements in any period on the periodic table. However, the screening increases toward the bottom of any family.

The effective nuclear charge is useful in explaining most of the observed trends. For example, there is a trend of decreasing atomic radii when moving from left to right across any period on the periodic table. The atomic radius is the distance the valence electrons are from the nucleus. In going from top to bottom on the periodic table, each row adds an additional energy level. Thus, the valence electrons are farther from the nucleus and the atoms increase in size. (There are some minor exceptions, especially in the transition metals.) The atomic radius decreases when moving from left to right within a period. The nuclear charge is increasing across the period.

The number of core electrons is constant, so the screening is constant. The effective nuclear charge is increasing due to the increase in nuclear charge minus the constant screening. This increase exerts a greater attractive force on the valence electrons. The increased attraction pulls the outer electrons closer to the nucleus, resulting in a smaller atom.

Ionization Energy

Neutral atoms may gain or lose electrons to form **ions**. Ions are atoms (or groups of atoms) that have an overall charge. We will only consider ions consisting of one atom in this chapter. Ions consisting of only one atom are **monatomic ions**. If electrons are lost, the resultant ion has more protons than electrons and therefore has a positive charge. An ion having a positive charge is a **cation**. The cation has one positive charge for each electron lost. How easily one or more electrons are lost depends on the atomic radius and the effective nuclear charge.

The **ionization energy** (IE) is the energy required to remove an electron from a gaseous atom in its lowest possible energy state, its ground state. This process begins with the electrons farthest from the nucleus. It requires energy to overcome the attractive force of the nucleus. The closer the electron is to the nucleus, the greater the attraction. The greater the attraction, the greater the energy required to remove that electron. Therefore, ionization energies tend to decrease going toward the bottom of a family. This is because the valence electrons are farther from the nucleus. This is the reverse of the atomic radii trend. Larger atoms have smaller ionization energies.

BTW

Atoms or groups of atoms may lose electrons to form positive ions, known as cations.

It is possible to remove more than one electron. This yields a second or a third IE, and so on. Successive ionization energies require more energy than the preceding one. This is true because the increase in positive charge creates a greater attraction for the remaining electrons, pulling those electrons closer to the nucleus. More energy is necessary to overcome this increased attraction.

The IE increases from left to right across a period. This is because the effective nuclear charge is increasing. This increase leads to a greater attraction, which requires more energy to overcome.

Increasing Ionization Energy →

IA	IIA	IIIB	IVB	VB	VIB	VIIB	VIIIB			IB	IIB	IIIA	IVA	VA	VIA	VIIA	VIIIA
1 H																	He
2 Li	Be											B	C	N	O	F	Ne
3 Na	Mg											Al	Si	P	S	Cl	Ar
4 K	Ca	Sc	Ti	V	Cr	Mn	Fe	Co	Ni	Cu	Zn	Ga	Ge	As	Se	Br	Kr
5 Rb	Sr	Y	Zr	Nb	Mo	Tc	Ru	Rh	Pd	Ag	Cd	In	Sn	Sb	Te	I	Xe
6 Cs	Ba	La	Hf	Ta	W	Re	Os	Ir	Pt	Au	Hg	Tl	Pb	Bi	Po	At	Rn
7 Fr	Rd	Ac															

Increasing Ionization Energy →

However, the trend is not linear. For example, in the second period there are peaks at beryllium, Be; nitrogen, N; and neon, Ne. These peaks in ionization energies correspond to very stable electron configurations of the atoms (very stable compared to the general trend). For example, the electron configuration of beryllium is $1s^2 2s^2$. It has a filled valence sublevel that provides additional stability. In nitrogen, the $2p$ sublevel is half-filled. The electrons are in different $2p$-orbitals. Being in different orbitals, they are as widely separated as possible. This separation minimizes the repulsion among the negative charges. This leads to an increase in stability. The increased stability leads to an increase in the IE. In neon, the second energy level is full, making the electron configuration particularly stable. High ionization energies occur for stable electron configurations. The stable electron configurations are filled shells and subshells, and half-filled subshells.

Ions, like atoms, have size. For ions, the term is **ionic radii**. For cations, the loss of electrons results in a decrease in size, since (for the representative metals) an entire energy level is usually lost. A sodium ion, Na^+, is smaller than a sodium atom. The greater the number of electrons removed, the greater the decrease in radius. This applies to any element and its cations as illustrated by the trend in radii of $Fe > Fe^{2+} > Fe^{3+}$.

Electron Affinity

It is possible to add one or more electrons to an atom. This process yields an ion with a negative charge. An ion with a negative charge is an **anion**. The energy change accompanying the addition of an electron to a gaseous atom in its ground state is the **electron affinity (EA)**. The first EA may be endothermic, exothermic, or even zero. The second EA is for the addition of a second electron. Each EA, after the first, requires more energy. This is because the approaching negative electron must overcome the repulsion of the negative ion charge. Electron affinities are less reliable than ionization energies. The reason for this is that EAs are more difficult to measure.

The general trend of electron affinities is like the trend for ionization energies. Small atoms with a high effective nuclear charge have high EAs. The EAs tend to increase going up a column. There is also an increase toward the right on the periodic table. However, electron configurations complicate this trend. The noble gases, such as He and Ne, have essentially no electron affinities. This is due to their filled valence shells having no room for the electron. (Noble gases do have ionization energies.) The alkaline earth metals, such as Be and Mg, with their filled s^2 configurations, have essentially no EA. Additional complications appear when we consider elements other than the representative elements.

BTW

Atoms may gain electrons to form monatomic anions.

The addition of one or more electrons influences the size of the ion formed. When an atom gains electrons, it produces an anion having a larger ionic radius than the original atom. The more electrons added, the greater the increase in size. For example, we see the following trend in radii with oxygen: $O < O^- < O^{2-}$. The effective nuclear charge is constant; however, the increasing number of electrons leads to an increase in repulsion. This pushes the electrons farther from each other and the nucleus.

Periodic Trends

Mendeleev first developed the periodic table by organizing the elements to emphasize trends in their chemical and physical properties. It is important that you be able to recognize these trends also. There are general trends, such as the fact that the atomic radii increase in size as you move down a family on the periodic table. There are "exceptions," such as the higher than expected ionization energy of nitrogen. In most cases, you will only need to know the general trends. However, there may be times where you will need to explain an exception. Most of the exceptions involve a filled shell or subshell or a half-filled subshell. These electron configurations are especially stable. The trends within the transition elements are not often as regular as the representative elements.

Many of the trends in this chapter depend upon the strength of attraction of the nucleus upon the electrons. There are two factors relating to this attraction. The first is the distance the electron is from the nucleus. The other is the effective nuclear charge. These two factors are interdependent. An example of this interdependence is the gradual decrease in the atomic radius within any period on the periodic table. This decrease is due to the increase in the effective nuclear charge as we move to the right across a period. While we can make predictions concerning trends simply by looking at the relative position of an element on the periodic table, any explanation of a relative value must involve the two factors of size and effective nuclear charge.

Trends for the elements may be either horizontal or vertical. The combination of these leads to diagonal relationships that increase either from the lower left to the upper right on the periodic table or from the upper right to the lower left. There are few trends that increase along another diagonal or that are only horizontal or vertical. You should remember that the noble gases are usually not included in some trends. In addition, hydrogen is an exception to many trends.

There are two general classes of questions concerning periodic properties. One type simply asks for ranking elements in either increasing or decreasing order. You should be careful not to accidentally put the values in reverse order. Simply locating the elements on the periodic table often allows answering this category of question. Exceptions are usually not the focus. The other type asks for an explanation of values. In this case, you will need to use atomic radius and the effective nuclear charge. In many cases, the electron configuration of the elements is involved. In this kind of question, position on the periodic table is often not very important. To answer this type of question often requires you to deal with one or more exceptions.

Another type of question may ask for a comparison of apparently unrelated information. For example: Rank the following in order of increasing radius: Ne, F^-, and Na^+. In this case, we are comparing ionic and atomic radii of elements from the second and third periods on the periodic table. If we consider the electron configuration of each of these species, we find they are all $1s^2 2s^2 2p^6$. Since the electron configurations are identical, the screenings of the core electrons are identical. If there were no additional factors, all three of the species would be identical in size. The cation, Na^+, has a higher attraction for the electrons due to its positive charge, which leads to an increase in the effective nuclear charge. For this reason, the cation will be the smallest of the three. The anion, F^-, has additional repulsion due to the extra electron indicated by the charge. This additional electron leads to an increase in repulsion of the electrons leading to an increase in size. The anion is the largest of the three species. Thus, the answer to the question is $Na^+ < Ne < F^-$.

There are other exceptions. For example, the EA of fluorine is lower than expected. This is because of the repulsion of the electrons in the small fluoride ion. If the ion were larger, the repulsion would be lower. The larger chlorine atom has the highest EA of all the elements.

EXERCISES

EXERCISE 8-1

Answer the following questions on periodic trends.

1. Define effective nuclear charge.

2. What term refers to a column on the periodic table?

3. How does the effective nuclear charge change among the members of a period on the periodic table?

4. True or False: The ionization energy may be either endothermic or exothermic.

5. True or False: The ionization energy is higher than expected for filled and half-filled subshells.

6. The products resulting from the action of the ionization energy on a gaseous atom are an electron and which of the following?

    a.   an anion
    b.   another atom
    c.   a cation
    d.   a polyatomic ion
    e.   an octet

7. How does the effective nuclear charge affect the ionization energies of the elements in a period on the periodic table?

8. How does the atomic radius affect the ionization energies of the elements in a family on the periodic table?

9. Which elements in the third period of the periodic table are likely to have unusually high ionization energies?

10. How does the ionic radius of K^+ compare to the atomic radius of K?

11. True or False: The electron affinity may be either endothermic or exothermic.

12. The product resulting from the action of the electron affinity and an electron on a gaseous atom are:
    a. an anion
    b. another atom
    c. a cation
    d. a polyatomic ion
    e. an octet

13. How does the effective nuclear charge affect the electron affinities of the elements in a period on the periodic table?

14. How does the atomic radius affect the electron affinities of the elements in a family on the periodic table?

15. How does the ionic radius of Cl^- compare to the atomic radius of Cl?

16. Rank the following elements in order of increasing ionization energy: Ar, K, and Br.

17. Rank the following elements in order of increasing electron affinity: Ar, K, and Br.

18. Rank the following in order of increasing radius: O^{2-}, Ne, F^-, Mg^{2+}, and Na^+.

19. How would the energy necessary to convert a gaseous Al atom to a gaseous Al^{3+} ion be found?

20. Why is the electron configuration of chromium, $1s^2 2s^2 2p^6 3s^2 3p^6 4s^1 3d^5$, an exception to the aufbau principle?

Flashcard App

Lewis Structures and Chemical Bonding

MUST KNOW

- Lewis symbols and structures are a type of structural formula in which shared pairs of electrons are represented by dashes and unshared electrons are shown as dots.

- The Lewis structure may be generated using the $N - A = S$ rule.

- The electronegativity of an element is the measure of the attraction that an element has on the electrons in a chemical bond.

- In general, single bonds are longer and weaker than double bonds, while triple bonds are the shortest and strongest.

y the time we've gone through this chapter we'll have achieved an understanding of Lewis structures and chemical bonding, both ionic and covalent. You might need to review the section in Chapter 2 on chemical formulas. Chapter 6 on Hess's law may also be helpful. Ionization energies and electron affinities, from Chapter 8, are also important. Let's see what we've got.

Lewis Symbols

Chemical compounds are pure substances composed of atoms in specific ratios held together by chemical bonds. The basic principle that governs bonding is the observed stability of the noble gas family (group 8A or group 18 elements). Their extreme stability is related to the fact that they have a filled valence shell. For all noble gases, except helium, this is a full complement of eight valence electrons. This is the basis for the **octet rule**. During chemical reactions, atoms tend to gain, lose, or share electrons to achieve an octet of electrons in their outer shell. By this process, the elements become isoelectronic with the closest noble gas. **Isoelectronic** means that the species have the same number and arrangement of electrons. There are exceptions to the octet rule, but most of the time it applies.

> **BTW**
> The octet rule doesn't always work, but for the representative elements, it works most of the time.

In later sections we will show you two types of chemical bonds: ionic and covalent. It is important to be able to represent compounds in terms of the atoms and valence electrons that make up the chemical species (compounds or polyatomic ions). One of the best ways is to use Lewis symbols and structures.

> **BTW**
> Here, as in most cases, hydrogen is an exception. Hydrogen achieves stability with a pair instead of an octet.

The **Lewis symbol**, or Lewis electron-dot symbol, is a way of representing an element and its valence electrons. To create a Lewis symbol, we begin by writing the element's chemical symbol. This represents the atom's nucleus and all core electrons. We then add symbols, usually dots, indicating the valence electrons around the atom's symbol. We distribute the valence electron dots one at a time around the symbol, placing the dots on either side, above, or below the symbol. This continues until, if necessary, there are four separate electrons present. If there are more than four electrons, you will need to pair the electrons until you can account for all the valence electrons. The following figure shows the Lewis symbol for several different elements. We will be using Lewis symbols extensively in the discussion of bonding, especially covalent bonding.

Lewis Symbols for Five Common Elements

$$H \cdot \qquad \cdot \overset{\displaystyle \cdot}{\underset{\displaystyle \cdot}{C}} \cdot \qquad \cdot \overset{\displaystyle \cdot \cdot}{\underset{\displaystyle \cdot}{N}} \cdot \qquad \cdot \overset{\displaystyle \cdot \cdot}{\underset{\displaystyle \cdot \cdot}{O}} : \qquad \cdot \overset{\displaystyle \cdot \cdot}{\underset{\displaystyle \cdot \cdot}{Cl}} :$$

In Chapter 2, we saw molecular and empirical formulas. Recall that the **molecular formula** indicates the kind and actual number of atoms present. The **empirical formula** simply shows the kind of atoms present and their lowest whole number ratio. In this chapter, structural formulas are important. The Lewis formula of a compound is an example of a structural formula. A **structural formula** shows the number and type of atoms present, as well as the bonding pattern. The bonding pattern shows which atoms bond to each other. It usually does not give you an indication of the actual shape of the molecule. (For a discussion on the shape of molecules, see Chapter 10 on Molecular Geometry and Hybridization.) The following are the empirical, molecular, and Lewis structural formulas for hydrogen peroxide, H_2O_2.

BTW

For the representative elements, the valence electrons are all electrons in the outer s- and p-orbitals of an atom. A quick way of determining the number of valence electrons is to locate the element on the periodic table. There are eight columns of representative elements. The first column, headed by H and Li, has one valence electron, and the second column has two. Skipping the transition elements, the next column, headed by B and Al, has three. This continues to the last (eighth) column where there are eight valence electrons. The only exception to this procedure is helium, which has only two valence electrons.

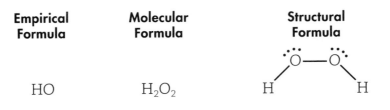

Empirical Formula	Molecular Formula	Structural Formula
HO	H_2O_2	

IRL Neither the empirical nor molecular formula is definitive in identifying a specific compound; generally, it takes the structural formula to identify a specific compound. Both ethyl alcohol (the drinking kind—C_2H_5OH) and dimethyl ether (a highly flammable substance—CH_3OCH_3) have the same empirical and molecular formulas, but they certainly don't have the same structural formula or properties.

Electronegativity

In many cases, you will need to predict the type of bond present. We will be giving you several tips along the way to help you decide. Electronegativity is a tool that you can use in predicting bond type. This tool will also help you to make other predictions.

Electronegativity is a measure of the attractive force that an atom in a compound exerts on electrons in a bond. You may find electronegativity values in a table in your textbook. There are two trends in these values. In general, electronegativities increase going toward the right on the periodic table (excluding the noble gases). The values also increase toward the top. These trends combine and lead to fluorine being the most electronegative element. The nearer an element is to fluorine, the higher its electronegativity.

You will need to know the electronegativity difference between the atoms. The key to the type of bond formed is this difference. A mostly ionic bond is present when there is a large electronegativity difference. "Large," according to Linus Pauling, refers to a difference greater than 1.7. If the difference is zero, the bond is covalent. Anything in between these extremes is polar covalent. (Later in this chapter we will discuss these bond types.) You do not need to know the actual electronegativity values (you can find them on the internet should you choose). The farther apart two elements appear on the periodic table, the greater the electronegativity difference. Widely separated elements usually form ionic bonds.

> **BTW**
>
> *If the atoms are widely separated on the periodic table (metals and nonmetals), the bonding is most likely ionic. Nonmetals close together are likely to form covalent bonds.*

Ionic Bonds and Lattice Energy

Ionic bonding involves the transfer of electrons from one atom to another. The more electronegative element gains electrons. The less electronegative element loses electrons. This results in the formation of **cations** and **anions**. Usually, an ionic bond forms between a metal and a nonmetal. The metal loses electrons to form a cation. The nonmetal gains electrons to become an anion. The attraction of the opposite charges forms an ionic solid.

> **BTW**
>
> *Cations have a positive charge and anions have a negative charge.*

The number of electrons lost or gained depends on the number of electrons necessary to lead to an octet. In general, an atom can gain or lose one or two and on rare occasions three electrons, but not more than that. Potassium, K, has one valence electron in energy level 4. If it loses that one, it only has three filled shells remaining. The "new" outer shell has an octet. Bromine, Br, has seven valence electrons, so if it gains one electron it will have its octet. A chemical reaction takes place between the potassium and bromine with an electron moving from the potassium atom to the bromine atom— completing the octet for both. In this manner, the ionic compound potassium bromide, KBr, forms.

> **BTW**
>
> *Elements on the left side of the periodic table (the metals) react with elements on the right side (the nonmetals) to form ionic compounds (salts).*

The reaction of magnesium, with two valence electrons, and chlorine, with seven valence electrons, will produce magnesium chloride. The magnesium must donate one valence electron to each of *two* chlorine atoms. This leaves a magnesium ion and two chloride ions. All the ions have a complete octet. The ions form the ionic compound magnesium chloride, $MgCl_2$.

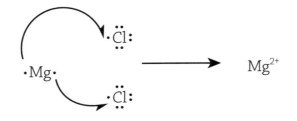

The total number of electrons lost must equal the total number of electrons gained. The reaction of aluminum with oxygen produces aluminum oxide. The aluminum has three valence electrons to lose. The oxygen has six valence electrons and needs two. The lowest common multiple between 3 and 2 is 6. It requires two aluminum atoms, losing three electrons each, to supply six electrons. It will require three oxygen atoms, gaining two electrons each, to account for the six electrons. The resultant compound, aluminum oxide, has the formula Al_2O_3. All ions present have an octet of electrons.

The formation of ions requires energy. The reaction cannot occur if insufficient energy is available. The source of the necessary energy is the lattice energy. The **lattice energy** is defined as the energy required to separate the ions in one mole of an ionic solid. There are ways of estimating the lattice energy, but we often find lattice energies by applying Hess's law. You learned how to apply Hess's law in Chapter 6. A Born–Haber cycle represents the application of Hess's law to the determination of lattice energies.

Covalent Bonds

Covalent bonding is the sharing of one or more *pairs* of electrons by two atoms. The covalent bonds in a **molecule**; a covalently bonded compound, are each represented by a dash. Each dash is a shared *pair* of electrons. These covalent bonds may be single bonds, one pair of shared electrons as in H-H; double bonds, two shared pairs of electrons as in $H_2C=CH_2$; or triple bonds, three shared pairs of electrons, $:N\equiv N:$. It is the same driving force to form a covalent bond as an ionic bond—completion of the atom's octet. In the case of the covalent bond, the sharing of electrons leads to both atoms using the electrons toward their octet.

In the hydrogen molecule, H_2, the atoms share the electrons equally. Each hydrogen nucleus has one proton equally attracting the bonding pair of electrons. A bond like this is a **nonpolar covalent bond**, or simply a covalent bond. In cases where the two atoms involved in the covalent bond are not the same, then the attraction is not equal. The bonding electrons are pulled more toward the atom with the greater attraction (more electronegative atom). This bond is a **polar covalent bond**. The atom that has the greater attraction takes on a partial negative charge and the other atom a partial positive charge.

Relative electronegativity values are important here. It is important to know which element has the greater attraction for the electrons. This is the atom with the greater electronegativity. Consider, for example, hydrogen fluoride, HF. The fluorine has a greater attraction for the bonding pair of electrons. For this reason, it takes on a partial negative charge. This will leave the hydrogen with a partial positive charge. The presence of partial charges gives a polar covalent bond. Many times, we use an arrow, in place of a dash, to represent this type of bond. The head of the arrow points toward the atom that has the greater attraction for the electron pair:

$$^{\delta+}\text{H}\longrightarrow\text{F}^{\delta-}$$
$$\longmapsto$$

The charges formed are not full charges as found in ions. These are only partial charges, indicated by a delta, δ. Many times, these polar bonds are responsible for the entire molecule being polar. The molecule is polar if it has a negative end and a positive end. Polar molecules attract other polar molecules, and this attraction may greatly influence the properties of that substance. (We will see the consequences of this in Chapter 11.)

BTW

Normally the atom needing the greater number of electrons to achieve an octet is the central atom. If two atoms need the same number of electrons, the larger is usually the central atom.

A Lewis structure can show the bonding pattern in a covalent compound. In Lewis formulas, we show the valence electrons that are not involved in bonding as dots surrounding the element symbols (though other symbols may be used). The valence electrons involved in bonding are present as dashes. There are several ways of deriving the Lewis structure, but here is one that works well for most compounds that obey the octet rule.

EXAMPLE

▶ Draw the Lewis structural formula for CH_3OH.

▶ First, write a general framework for the molecule. In this case, the carbon bonds to the oxygen since hydrogen can only form one bond. The way the formula is written, there are three hydrogen atoms attached to the carbon atom and one hydrogen atom attached to the oxygen atom.

BTW

Hydrogen is never the central atom.

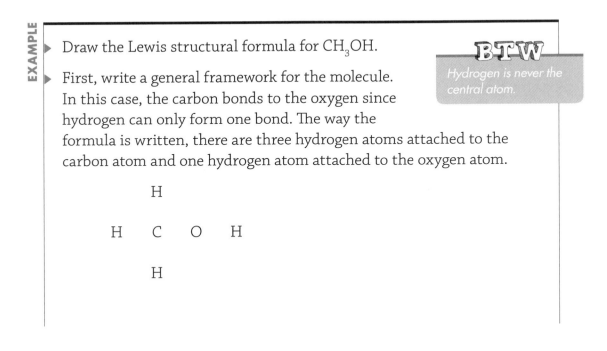

▶ To determine where to place the electrons, we will apply the $N - A = S$ rule, where:

> N = sum of valence electrons needed for each atom. The two allowed values are 2 for hydrogen and 8 for all other elements.
>
> A = sum of all available valence electrons
>
> S = the number of electrons shared and $S/2$ = the number of bonds

▶ For CH_3OH, we would have:

	1 C		**4 H**		**1 O**	
N	8	+	4 (2) =8	+	8	= 24
A	4	+	4 (1) =4	+	6	= 14

$$S = N - A = 24 - 14 = 10 \qquad \textbf{bonds} = S/2 = 10/2 = 5$$

▶ The result, 5, indicates that there are five bonds present in this compound. You should begin by placing one bond (pair of electrons) between each pair of atoms. Use a dash to represent a bond. If you still have bonds remaining after placing a dash between each pair of atoms, it means you have double or triple bonds. Next, distribute the remaining available electrons around the atoms so that each atom has its full octet, eight electrons (either bonding or nonbonding, shared or not). Hydrogen is an exception; it only gets two. The following figure shows the Lewis structural formula of CH_3OH.

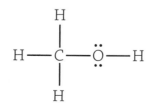

> ▶ Confirm for yourself that the number of electrons present equals the A in the table for the $S = N - A$ equation. If the number does not equal A, there is a problem.

It is also possible to write a Lewis structure for polyatomic anions or cations. The $N - A = S$ rule can be used, but for an anion, extra electrons equal to the magnitude of the negative charge must be added to the electrons available. If the ion is a cation, you will need to subtract the number of electrons equal to the charge.

Resonance

Sometimes, when writing the Lewis structure of a species, we may draw more than one possible "correct" Lewis structure for a molecule. The nitrate ion, NO_3^-, is a good example. The structures that we write for this polyatomic anion differ in which oxygen has a double bond to the nitrogen. None of these three truly represents the actual structure of the nitrate ion—it is an average of all three of these Lewis structures. We use resonance theory to describe this situation.

Resonance occurs when more than one Lewis structure (without moving atoms) is possible for a molecule. The individual structures are called resonance structures (or forms) and are written with a two-headed arrow (⟷) between them. The three resonance forms of the nitrate ion are as follows:

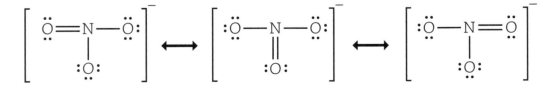

Again, let us emphasize that the actual structure of the nitrate ion is *not* any of the three shown. It is *not* flipping back and forth among the three. It is an average of all three. All the bonds are the same and are intermediate between single bonds and double bonds in strength and length.

When you draw resonance structures, you can only move electrons (bonds). Never move the atoms.

Bond Energy and Bond Length

When writing the Lewis structures of polyatomic ions, don't forget to show their ionic charge.

The reason bonds form is that bonded species are more stable than nonbonded species. Since bonds are stabilizing factors, it requires energy to break a bond. The **bond energy** is the energy required to break a bond. Since bond energy is the energy required, the values are always positive (endothermic). We also know that if energy is necessary to break a bond, the reverse process, the formation of a bond releases an equivalent amount of energy.

Bond energies, found in tables such as in a reference book or on the internet, are average values. In a molecule, the actual bond energies may be larger or smaller than the average. The values will always be greater than the average if resonance is present. The strength of a bond depends upon the identity of the atoms involved and the type of bond, increasing in the order: single, double, or triple bond.

When we use bond energies to estimate the heat of reaction, we need to have the values for all the bonds broken and for all the bonds formed. The bonds broken are all the bonds in the reactants, and the bonds formed are all the bonds in the products. The difference between these energy totals is the heat of reaction:

$$\Delta H = \Sigma \text{ bonds broken} - \Sigma \text{ bonds formed}$$

The **bond length** is the distance between the two atomic nuclei of the atoms involved in the bond. The bond length is an average distance and varies with the atoms involved and the type of bond, decreasing in the order:

single, double, or triple bond. It is important to realize that a double bond is not twice the strength or half the length of a single bond. Without looking at tabulated values, we can simply say that double bonds are stronger and shorter than single bonds.

More About Lewis Structures

Let's work two examples illustrating the steps necessary to produce a good Lewis structure. We will take HNO_2 and XeF_4 for these examples.

▸ The first molecule is nitrous acid. It is an example of an oxyacid. (Oxyacids are compounds containing hydrogen, oxygen, and one other element.) The other compound is xenon tetrafluoride.

▸ There is more than one way to arrange the atoms in nitrous acid. However, only one will lead to a good Lewis structure. You should avoid any arrangement that places identical atoms adjacent. Identical atoms, other than carbon, rarely bond to each other unless there is no alternative. This is an oxyacid. The hydrogen is acidic. When you arrange the atoms for an oxyacid, an acidic hydrogen atom will always attach to an oxygen atom. This oxygen atom will have only one hydrogen atom attached. This oxygen will also attach to the other element in the compound. If there is more than one acidic hydrogen atom, attach one to a separate oxygen atom. The resultant order of atoms that you get might be HONO. It does not matter to which oxygen you attach the hydrogen atom. The arrangement ONOH is the same as HONO. You may arrange these atoms in a horizontal arrangement as done here, or a vertical arrangement, or any other

When you begin a Lewis structure, do not place identical atoms adjacent to each other unless there is no alternative. Carbon is the only common exception. Most compounds containing more than one carbon atom will have the carbon atoms adjacent to each other.

way. It makes no difference. We will use the first arrangement in this example.

▶ We know that there needs to be at least one bond between each pair of atoms. We also know that the bond from hydrogen to the adjacent oxygen will be the only bond to the hydrogen atom. This information should be on your mind as you move on to complete the structure.

▶ If you wish to apply the $S = N - A$ rule, you will need to assign values to N and A. To find N, we use 2 for hydrogen and 8 for each of the other three atoms. This gives $N = 2 + 3(8) = 26$. The determination of A uses 1 for hydrogen, 5 for nitrogen, and 6 for each oxygen atom. This gives $S = 1 + 5 + 2(6) = 18$. Using the values for N and A, we find $S = 26 - 18 = 8$. If $S = 8$, then there are $S/2 = 4$ bonds.

▶ We can place one bond between each pair of atoms to get H-O-N-O. This arrangement accounts for three of the four bonds. Since there is one more bond to account for, we need to create a double bond. The double bond cannot involve the hydrogen, since it already has the one and only bond that it can have. The oxygen next to the hydrogen is not a likely candidate, since it already has two bonds. The most likely candidate for another bond is the other oxygen atom. We can add this bond, which changes the N-O bond to a double bond. The bonding arrangement is now H-O-N=O. The four bonds, at two electrons each, account for eight of the available electrons (A). To finish the structure, we need to add the remaining 10 electrons ($A - 8$). Usually electrons come in pairs, so our 10 electrons will appear as five pairs.

▶ We will begin with the oxygen atoms since oxygen is the most electronegative element in this compound. (The most electronegative elements *will* get its octet.) Each oxygen atom has two bonds (four electrons), so each atom needs four electrons (two pairs). We need to add two separate pairs to each oxygen atom. It is not too important where you place the pairs if you do not place the electrons between the oxygen and another atom. These pairs account for eight more electrons. The remaining two electrons constitute one more pair. This pair will probably go on the nitrogen, but you should check to make sure. The nitrogen atom has three bonds (six electrons), and it needs two more to achieve an octet. Therefore, we can add our last two electrons as a pair on the nitrogen. The positioning of the pair is not important. Do not add the two electrons separately.

▶ The final Lewis structure for nitrous acid follows:

$$H - \ddot{O} - \ddot{N} = \ddot{O}$$

▶ In this structure, we have two central atoms. The nitrogen is one, and the oxygen with the hydrogen is the other.

EXTRA HELP

EXAMPLE

▶ Now we will begin drawing the Lewis structure of XeF_4. Xenon will be the central atom, and we will arrange fluorine atoms around it. In this way, we avoid attaching identical atoms to each other. We will need a bond between the central xenon and each of the fluorine atoms. This arrangement means there will be at least four bonds.

▶ If you wish to apply the $S = N - A$ rule, you will need to assign values to N and A. To find N, we use 8 for xenon and 8 for each of the fluorine atoms. This gives $N = 8 + 4(8) = 40$. The determination of A uses 8 for xenon and 7 for each fluorine atom. This gives $A = 8 + 4(7) = 36$. Using the values for N and A, we find $S = 40 - 36 = 4$. If $S = 4$, then

BTW

All the available electrons, A, must appear in the final Lewis structure. They will normally appear in pairs unless there is an odd number of electrons. If the number of electrons is odd, there will be only one unpaired electron.

there are $S/2 = 2$ bonds. This is a problem since we already know we need at least four bonds. This means that we have a compound that is an exception to the octet rule.

▸ Even though we have an exception, we can still complete the Lewis structure. We need to draw a bond from each of the fluorine atoms to the central xenon. This gives us four bonds and uses eight electrons. Each fluorine atom needs to complete its octet. The bond accounts for two electrons, so we need six more electrons (three pairs) for each. Therefore, we add three separate pairs to each of the fluorine atoms. Six electrons per fluorine times four fluorine atoms accounts for 24 electrons. Our Lewis structure now contains $8 + 24 = 32$ electrons. The number of available electrons (A) is 36, so we still need to add $36 - 32 = 4$ electrons. These four electrons will give us two pairs. The xenon atom will get these pairs and become an exception to the octet rule. The actual placement of the pairs is not important if it is obvious that they are with the central atom and not one of the fluorine atoms. The final Lewis structure follows:

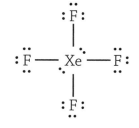

In a compound that is an exception to the octet rule, there is usually only one atom, other than hydrogen, that is an exception. There are few nonhydrogen compounds with more than one exception present.

Elements in the first two rows on the periodic table will never exceed an octet. If they are exceptions to the octet rule, they will have less than an octet.

When you are not certain where to put electrons, remember that the most electronegative element will get its octet. This element will not exceed an octet.

EXERCISES

EXERCISE 9-1

Answer the following questions.

1. State the octet rule.

2. What does *isoelectronic* mean?

3. How many electrons should the Lewis symbols of each of the following elements indicate?
   a. K
   b. Ar
   c. I
   d. N
   e. Si

4. What is the definition of electronegativity?

5. Which of the following pairs of elements will have the greatest difference in electronegativity?
   a. B and C
   b. C and Si
   c. O and F
   d. Br and S
   e. Cl and Si

6. Which of the following is not following the octet rule?
   a. F^-
   b. O^-
   c. P^{3-}
   d. Mg^{2+}
   e. Ti^{4+}

7. How many electrons must each of the following gain to achieve an octet?
    a. S
    b. I
    c. Ar

8. How many electrons must each of the following lose to achieve an octet?
    a   Cs
    b. Mg
    c. Cl

9. What is the definition of lattice energy?

10. What is the maximum number of covalent bonds between a pair of atoms?
    a. 1
    b. 2
    c. 4
    d. 3
    e. 6

11. Which type of bond will form between two nonmetals that differ slightly in electronegativity?
    a. polar covalent
    b. nonpolar covalent
    c. ionic
    d. metallic
    e. no bond can form

12. Which of the following atoms can never be the central atom in a Lewis structure?
    a. C
    b. H
    c. O
    d. N
    e. B

13. Arrange the following in order of decreasing bond length: N-N, N≡N, and N=N.

14. Arrange the following in order of increasing bond energy: C-O, C≡O, and C=O.

15. Using Lewis symbols, write a balanced chemical equation showing the formation of lithium fluoride, LiF, from isolated lithium and fluorine atoms.

16. Using Lewis symbols, write a balanced chemical equation showing the formation of calcium fluoride, CaF_2, from isolated calcium and fluorine atoms.

17. Using Lewis symbols, write a balanced chemical equation showing the formation of oxygen difluoride, OF_2, from isolated oxygen and fluorine atoms.

18. Draw the Lewis structure of hypochlorous acid, HOCl.

19. Draw the Lewis structure of sodium phosphate, Na_3PO_4. (Be careful! This is tricky.)

EXERCISE 9-2

Draw the Lewis structure of each of the following.

1. H_2S

2. CO_2

3. Cl_2O

4. NH_4^+

EXERCISE 9-3

Draw the Lewis structure of each of the following.

1. BF_3

2. SF_4

3. XeF_2

4. CS_2

5. SF_6

EXERCISE 9-4

Draw the resonance structures of each of the following.

1. NO_2^-

2. SO_3

3. NO_2

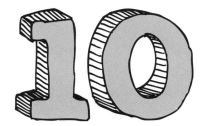

Molecular Geometry and Hybridization

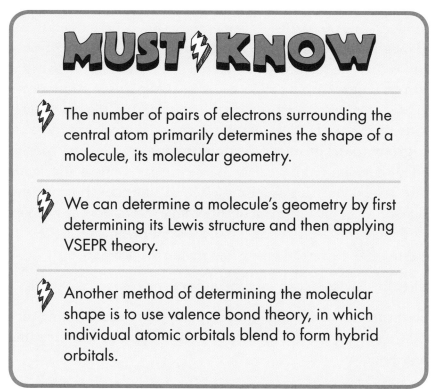

MUST ⚡ KNOW

⚡ The number of pairs of electrons surrounding the central atom primarily determines the shape of a molecule, its molecular geometry.

⚡ We can determine a molecule's geometry by first determining its Lewis structure and then applying VSEPR theory.

⚡ Another method of determining the molecular shape is to use valence bond theory, in which individual atomic orbitals blend to form hybrid orbitals.

e're now going to delve into electron and molecular geometry and hybridization. You might want to review the section in Chapter 7 on electron configuration. The section in Chapter 9 on writing Lewis structures is also important. Let's get going.

Molecular Geometry (VSEPR)

In Chapter 9, we indicated that the Lewis structure showed the bonding pattern of a molecule, but not necessarily its shape. The shape of a molecule greatly influences its properties. One method to predict the shape of molecules is the **valence-shell electron pair repulsion (VSEPR) theory**. The basis of this theory is that the valence-shell electron pairs around a central atom will try to move as far away from each other as possible. This includes electrons in bonds and elsewhere. The electrons do this to minimize the repulsion between the like (negative) charges. We will determine two geometries. The first is the **electron-group (pair) geometry**. The electron-group geometry considers all electron pairs surrounding a nucleus. The second is the **molecular geometry**. In this case, the nonbonding electrons (electron lone pairs) become "invisible." We consider only the arrangement of the atomic nuclei. For the purposes of geometry, double and triple bonds count the same as single bonds.

To determine the electron-group and molecular geometry:

- Write the Lewis electron-dot formula of the compound.

- Determine the number of electron pair groups surrounding the central atom(s). Remember that double and triple bonds count the same as a single bond.

- Determine the geometric shape that maximizes the distance between the electron groups. This is the electron-group geometry.

- Mentally allow the nonbonding electrons to become invisible. They are still present and are still repelling the other electron pairs. However, we just don't "see" them. We then determine the molecular geometry from the arrangement of bonding pairs around the central atom.

The following diagram shows the electron-group and molecular geometry for two to six electron pairs.

Electron-Group and Molecular Geometry

Total Electron Pairs	Electron-Group Geometry	Bonding Pairs	Nonbonding Pairs (Lone Pairs)	Molecular Geometry
2 pairs	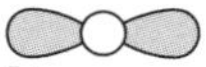 Linear	2	0	
3 pairs	 Trigonal planar	3	0	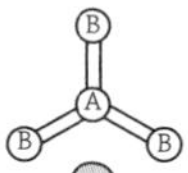
	Bent	2	1	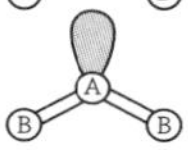
4 pairs	 Tetrahedral	4	0	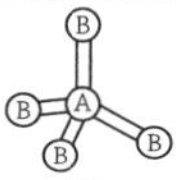
	Trigonal pyramid	3	1	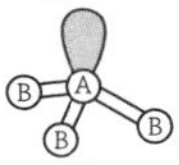
	Bent	2	2	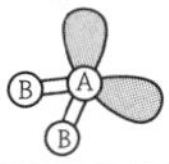
5 pairs	 Trigonal bipyramidal	5	0	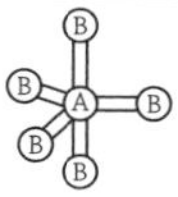
	Distorted tetrahedral (See-saw)	4	1	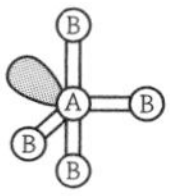
	T-shaped	3	2	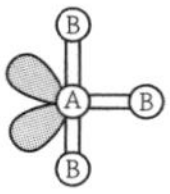
	Linear	2	3	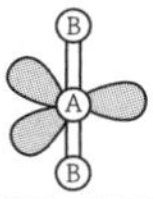
6 pairs	 Octahedral	6	0	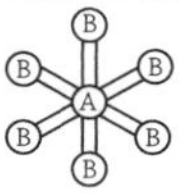
	Square pyramid	5	1	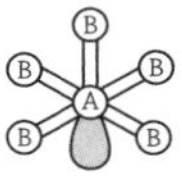
	Square planar	4	2	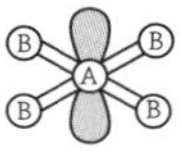

▶ For example, let's determine the electron-group and molecular geometry of carbon dioxide, CO_2, and water, H_2O. At first glance, you might imagine that the geometry of these two compounds would be similar since both have a central atom with two groups (atoms) attached. However, let's see if that is true.

▶ First, write the Lewis structure of each. These figures show the Lewis structures of carbon dioxide and water:

▶ Next, determine the electron group geometry of each. For carbon dioxide, there are two electron groups around the carbon (the two double bonds). Two electron pairs are a linear structure. For water, there are four electron pairs around the oxygen, two bonding and two nonbonding electron pairs. The presence of four total pairs gives tetrahedral electron-group geometry.

▶ Finally, mentally allow the nonbonding electron pairs to become invisible. What remains is the molecular geometry. For carbon dioxide, all groups are involved in bonding. There are no "invisible" groups. This means the electron group and the molecular geometry are the same (linear). However, water has two nonbonding pairs of electrons (now invisible). The remaining bonding electron pairs (and hydrogen nuclei) are in a bent molecular arrangement. The angle between these pairs is like that in the tetrahedral structure. The presence of lone pairs makes the angle slightly smaller than ideal.

Determining the molecular geometry of carbon dioxide and water also explains why their polarities are different. Carbon dioxide is not polar, and water is. This is true even though both are composed of polar covalent bonds. To be a polar molecule, one end must have a partial positive charge ($\delta+$) and the other must have a partial negative charge ($\delta-$). Carbon dioxide, because of its linear shape, has partial negative charges at both ends and a partial positive charge in the middle. Water, because of its bent shape, has a partial negative end, the oxygen, and a partial positive end, the hydrogen side. Carbon dioxide does not have a partial positive end. The polarity of the molecule is important because polar molecules attract other polar molecules. This may dramatically affect the properties of that substance. This is true with water. (See Chapter 11 for a discussion of water and intermolecular forces.)

BTW

In using the VSEPR theory to determine the molecular geometry, start first with the electron group geometry, make the nonbonding electrons mentally invisible, and then describe what remains.

IRL The molecular geometry that a molecule assumes is very important, especially in a biological system. For example, there are two types of glucose that have the same empirical, molecular, and simple structural formulas. They differ only in the arrangement in three-dimensional space of the atoms. One is biologically active and is used in intravenous solutions; the other form is totally inert in the body.

Valence Bond Theory (Hybridization)

VSEPR theory is one way to determine the molecular geometry. Another method involves using valence bond theory. **Valence bond theory** describes covalent bonding in terms of the blending of atomic orbitals to form new types of orbitals: hybrid orbitals. **Hybrid orbitals** are orbitals formed due to the combining of the atomic orbitals of the central atom. The total number of orbitals does not change. The number of hybrid orbitals equals the number of atomic orbitals used. The type of hybrid orbitals formed depends on the number and type of atomic orbitals used. The following table shows the hybrid orbitals resulting from the mixing of *s*-, *p*-, and *d*-orbitals. The atoms share electrons through the overlapping of their orbitals. Any combination of overlapping orbitals is acceptable.

Hybridization of *s*-, *p*-, and *d*-Orbitals

	Linear	Trigonal Planar	Tetrahedral	Trigonal Bipyramidal	Octahedral
Atomic orbitals mixed	one *s* one *p*	one *s* two *p*	one *s* three *p*	one *s* three *p* one *d*	one *s* three *p* two *d*
Hybrid orbitals formed	two *sp*	three *sp*^{2}	four *sp*^{3}	five *sp*^{3}*d*	six *sp*^{3}*d*^{2}
Unhybridized orbitals remaining	two *p*	one *p*	none	four *d*	three *d*
Orientation					

sp hybridization results from the overlap of one *s*-orbital with one *p*-orbital.	These two orbitals give two *sp* hybrid orbitals with a bond angle of 180°. This is a linear orientation.
sp² hybridization results from the overlap of one *s*-orbital with two *p*-orbitals.	Three *sp*^{2} hybrid orbitals form with a trigonal planar orientation and a bond angle of 120°. This type of bonding occurs in the formation of the C to C double bond as in $CH_2=CH_2$.
sp³ hybridization results from the mixing of one *s*-orbital and three *p*-orbitals, resulting in four *sp*^{3} hybrid orbitals with a tetrahedral geometric orientation.	We find this *sp*^{3} hybridization in carbon when it forms four single bonds with a bond angle of 109.5°.
sp³d hybridization results from the blending of one *s*-orbital, three *p*-orbitals, and one *d*-orbital.	The result is five *sp*^{3}*d* orbitals with a trigonal bipyramidal orientation, with bond angles of 90° and 120° (this is the only hybridization covered here with two different bond angles). This type of bonding occurs in compounds like PCl_5, an exception to the octet rule.
sp³d² hybridization occurs when one *s*, three *p*, and two *d*-orbitals mix, giving an octahedral arrangement. SF_6 is an example.	Again, this is an exception to the octet rule. If one starts with this structure and one of the bonding pairs becomes a lone pair, a square pyramidal shape results; two lone pairs give a square planar shape. The bond angle is 90°.

The following figure shows the hybridization that occurs in ethylene, $H_2C=CH_2$. Each carbon has sp^2 hybridization. On each carbon, two of the hybrid orbitals overlap with an s-orbital on a hydrogen atom to form a carbon-to-hydrogen covalent bond. The third sp^2 hybrid orbital overlaps with the sp^2 hybrid on the other carbon to form a carbon-to-carbon covalent bond. Note that each carbon has a remaining p-orbital that has not undergone hybridization. These are also overlapping above and below a line joining the carbons.

Hybridization in Ethylene $H_2C=CH_2$

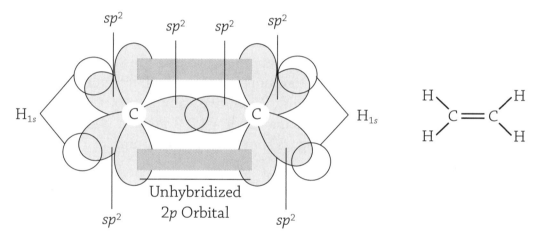

In ethylene, there are two types of bonds. **Sigma** (σ) bonds have the overlap of the orbitals on a line between the two atoms involved in the covalent bond. In ethylene, the C-H bonds and one of the C-C bonds are sigma bonds. **Pi** (π) **bonds** have the overlap of orbitals above and below a line through the two nuclei of the atoms involved in the bond. A double bond is always composed of one sigma and one pi bond. A carbon-to-carbon triple bond results from the overlap of one sp hybrid orbital and two p-orbitals on one carbon with the same on the other carbon. This results in one sigma bond (overlap of the sp hybrid orbitals) and two pi bonds (overlap of two sets of p-orbitals).

Molecular Orbital (MO) Theory

Another covalent bonding model is **molecular orbital (MO)** theory. In MO theory, atomic orbitals on the individual atoms combine to form molecular orbitals (MOs). These are not hybrid orbitals. An MO covers the entire molecule. Molecular orbitals have definite shapes and energies. The combination of two atomic orbitals produces two MOs. (The total number of orbitals never changes.) One of these MOs is a bonding MO. The other is an antibonding MO. The bonding MO has a lower energy than the original atomic orbitals. The antibonding MO has a higher energy. Lower energy orbitals are more stable than higher energy orbitals.

Once the MO forms, electrons enter. We add electrons using the same rules we used for electron configurations. The lower energy orbitals fill first (aufbau principle). There is a maximum of two electrons per orbital (Pauling exclusion principle). Orbitals of equal energy will half-fill orbitals before pairing electrons (Hund's rule). When two s atomic orbitals combine, two sigma (σ) MOs form. One is sigma bonding (σ). The other is sigma antibonding (σ^*). The following figure shows the MO diagram for H_2.

Molecular Orbital Diagram of H_2

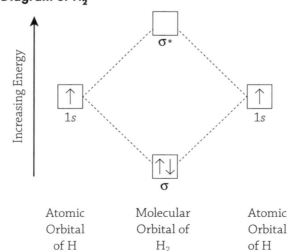

Note that the two electrons (one from each hydrogen atom) have both gone into the sigma bonding MO. We can determine the bonding situation in MO theory by calculating the MO bond order. The **MO bond order** is the number of electrons in bonding MOs minus the number of electrons in antibonding MOs, divided by 2. For H_2 in the preceding figure, the bond order would be $(2 - 0)/2 = 1$, which is a stable bonding situation that exists between two atoms when the bond order is greater than zero. The larger the bond order, the stronger the bond.

When two sets of p-orbitals combine, one sigma bonding and one sigma antibonding MO are formed along with two bonding (π) MOs and two pi antibonding (π^*) MOs. The following figure shows the MO diagram for O_2. For the sake of simplicity, the $1s$-orbitals of each oxygen and MOs have not been shown here, just the valence electron orbitals.

Molecular Orbital Diagram of Valence-Shell Electrons of O_2

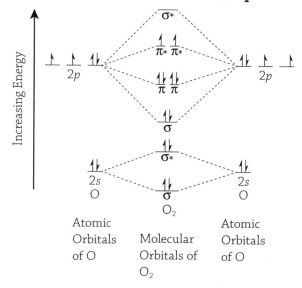

The bond order for O_2 would be $(10 - 6)/2 = 2$.

EXTRA HELP

Molecular Geometry Revisited

Let's examine the geometries of the two Lewis structures derived in the section "More About Lewis Structures" from the previous chapter.

$$H-\ddot{O}-\ddot{N}=\ddot{O} \qquad :\ddot{F}-\overset{\displaystyle :\ddot{F}:}{\underset{\displaystyle :\ddot{F}:}{Xe}}-\ddot{F}:$$

To predict the molecular geometries, you must have a correct Lewis structure. We will begin with the structure of nitrous acid. There are two central atoms in this structure: the nitrogen and the oxygen atom with the hydrogen attached; therefore, there will be two answers. We will begin with the nitrogen atom. This atom has an octet of electrons surrounding it. It has a lone pair and three bonding pairs distributed between one single and two double bonds. To predict the electron-group geometry, we count one for the lone pair, one for the single bond, and one for the double bond. This gives a total of three. Three pairs give a trigonal planar electron-group geometry, and they also imply a sp^2 hybridization. A trigonal planar electron-group geometry with one lone pair leaves us with a bent molecular geometry. As a bent species, with atoms of differing electronegativities, it is polar.

If we examine the other central atom, the oxygen with the attached hydrogen, we observe the presence of two lone pairs and two bonding pairs. The presence of these pairs and bonds, which total four, means that the electron-group geometry is tetrahedral. This arrangement has sp^3 hybridization. Since there are two lone pairs, the molecular geometry is bent.

An examination of the xenon tetrafluoride structure shows that the central atom, the xenon, has four bonds and two lone pairs. Six groups lead to an octahedral electron-group geometry. In this case, the hybridization is sp^3d^2. There are two lone pairs, so the molecular geometry is square planar. This geometry indicates that even though the bonds are polar covalent, their arrangement leads to a nonpolar molecule.

EXERCISES

EXERCISE 10-1

Answer these questions about molecular geometry and hybridization.

1. What does the abbreviation *VSEPR* stand for?

2. Reconstruct the first figure.

3. Determine the molecular geometries about the central atom for each of the following.
   a. H_2S
   b. Cl_2O
   c. BF_3

4. Determine the molecular geometries about the central atom for each of the following.
   a. SF_4
   b. XeF_2
   c. NH_4^+

5. Determine the molecular geometries about the central atom for each of the following.
   a. SF_6
   b. CS_2

6. Determine the hybridization of the central atom for each of the following.
   a. H_2S
   b. Cl_2O
   c. BF_3

7. Determine the hybridization of the central atom for each of the following.
    a. SF_4
    b. XeF_2
    c. NH_4^+

8. Determine the hybridization of the central atom for each of the following.
    a. SF_6
    b. CS_2

9. Which of the following molecules are polar?
    a. H_2S
    b. Cl_2O
    c. BF_3

10. Which of the following molecules are polar?
    a. SF_4
    b. XeF_2

11. Which of the following molecules are polar?
    a. SF_6
    b. CS_2

Flashcard App

Solids, Liquids, and Intermolecular Forces

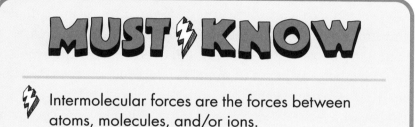

We are now going to examine intermolecular forces. Intermolecular forces are interactions between atoms, molecules, and/or ions. We can use these forces to explain both macroscopic and microscopic properties of matter.

The state of matter in which a substance exists depends upon two factors. One of these factors is the kinetic energy of the particles. The other factor is the intermolecular forces between the particles. The average kinetic energy of the molecules relates directly to the temperature. Kinetic energy tends to move particles away from each other. As the temperature increases, the average kinetic energy increases. When the kinetic energy increases, the particles move faster. In a solid, the intermolecular forces are sufficiently great to minimize particle movement. As the temperature increases, the kinetic energy increases and eventually will overcome the intermolecular forces. The substance will begin to **melt**. While a substance is melting, the temperature remains constant even though we are adding energy. The temperature at which the solid converts into the liquid state is the **melting point (mp)**. Melting is an example of a change in state. Changes of state, called **phase changes**, depend on the temperature. In some cases, the pressure can also influence these changes.

For now, we will concentrate on the solid and liquid states. You might want to review the section in Chapter 10 that deals with the polarity of molecules.

> **BTW**
> The melting point of a substance and its freezing point (fp) are identical.

Types of Intermolecular Forces

Intermolecular forces are attractive or repulsive forces between atoms, molecules, and/or ions. The attractive forces compete with the randomizing factor of kinetic energy. The structure that a substance exhibits depends on the strength and type of intermolecular forces present. Before we begin examining the different types of intermolecular forces, you may wish to refresh yourself by skimming Chapter 9 on bonding. It is important to recognize which molecules are polar and which are not. Polar molecules have partial positive and partial negative ends. Polar molecules have a dipole. These dipoles are one of the major factors involved in the intermolecular forces.

Covalent, metallic, and ionic bonds are very strong interactions. Some people consider these to be intermolecular forces. The following are weaker intermolecular forces. They appear in approximate order of decreasing strength. Even though weaker than bonds, they are nonetheless important.

Ion-Dipole Intermolecular Forces

This attraction is due to the attraction of an ion (cation or anion) with one end of a polar molecule (dipole). This type of attraction is especially important in aqueous salt solutions where the ion attracts water molecules and may form a hydrated ion (i.e., $Al(H_2O)_6^{3+}$).

BTW

The ion-dipole intermolecular force is different in that two different species must be present: an ion from one species and a polar molecule from a different species. These are very important in aqueous solutions.

Dipole-Dipole Intermolecular Forces

This attraction occurs between two polar molecules. It results from the attraction of the positive end of one dipole to the negative end of another dipole. For example, gaseous hydrogen chloride, $HCl(g)$, has a dipole. The hydrogen end has a partial positive charge and the chlorine end has a partial negative charge. The chlorine is more electronegative, so it has the partial negative charge. The partial-positive end of one HCl molecule attracts the partial-negative end of another HCl molecule. Dipole-dipole attractions are especially important in polar liquids. These dipole-dipole attractions tend to be rather strong intermolecular forces, although not nearly as strong as ion-dipole attractions.

Hydrogen Bonding Intermolecular Forces

Hydrogen bonding is really a subtype of dipole-dipole attraction. In this case, a hydrogen atom bonds to a very electronegative element. The only elements sufficiently electronegative are N, O, and F. The resultant bond to hydrogen is more polar than the electronegativity difference would predict. This extreme polarity leads to a greater than expected degree of charge separation. Therefore, the attraction of the hydrogen (bonded to N, O, or F) of one molecule and the N, O, or F of another molecule is unusually strong. Hydrogen bonds tend to be stronger than the typical dipole-dipole interaction, which means this type of intermolecular force is out of order on this list.

BTW

Hydrogen bonding can only occur when a hydrogen atom is bonded directly to an N, O, or F.

Hydrogen bonding explains why water has such unusual properties. We will discuss these properties later in this chapter.

Ion-Induced Dipole and Dipole-Induced Dipole Intermolecular Forces

These types of attractions occur when the charge on an ion or a dipole distorts the electron cloud of a nonpolar molecule. This induces a temporary dipole in the nonpolar molecule. These are relatively weak interactions. Like an ion-dipole force, this type of force requires the presence of two different substances.

London (Dispersion) Intermolecular Forces

This intermolecular attraction occurs in all substances. However, it is usually only significant for nonpolar substances. It arises from the momentary distortion of the electron cloud. This distortion causes a very weak temporary dipole, which induces a weak dipole in another molecule. These weak dipoles lead to an attraction. Although this is an extremely weak interaction, it is strong enough to let us liquefy nonpolar gases such as hydrogen, H_2. If there were no intermolecular forces attracting these molecules, it would be impossible to liquefy hydrogen. The more electrons present, the greater the London force.

Properties of Liquids

At the microscopic level, the liquid particles are in constant motion. The particles may exhibit short-range areas of order, but these usually do not last very long. Clumps of particles may form and then break apart. At the macroscopic level, a liquid has a specific volume but no fixed shape. Three additional macroscopic properties deserve discussion: surface tension, viscosity, and capillary action.

In the body of a liquid, intermolecular forces pull the molecules in all directions. At the surface of the liquid, the molecules pull down into the body of the liquid and from the sides. There are no molecules above the surface to pull in that direction. The effect of this unequal attraction is that the liquid tries to minimize its surface area. The minimum surface area for a given quantity of matter is a sphere. In a large pool of liquid, where sphere formation is not possible, the surface behaves as if it had a thin stretched elastic membrane or "skin" over it. The **surface tension** is the resistance of a liquid to an increase in its surface area. It requires force to break the attractive forces at the surface. The greater the intermolecular force, the greater the surface tension. Polar liquids, especially those that use hydrogen bonding, have a much higher surface tension than nonpolar liquids.

IRL   Surface tension is the reason a bug can walk across a pond or why you can float a steel paperclip on the surface of a glass of water.

Viscosity is the resistance to flow. Important factors influencing the viscosity of a liquid are the intermolecular forces and the temperature. The stronger the intermolecular force, the greater the viscosity. As the temperature increases, the kinetic energy of the particles increases. The higher kinetic energy will overcome the intermolecular attractive forces. This causes a lower viscosity. In some cases, another factor is the size of the molecule. Large and complex molecules will have difficulty moving past one another. If they cannot easily move past each other, the viscosity will be high.

Capillary action is the spontaneous rising of a liquid through a narrow tube against the force of gravity. It is due to competition of intermolecular forces within the liquid and attractive forces between the liquid and the tube wall. The stronger the attraction between the liquid and the wall, the higher the level rises. Liquids that have weak attractions to the walls, like mercury in a glass tube, have low capillary action. Liquids like water in a glass tube have strong attractions to the walls and will have high capillary action. This also explains why we observe a meniscus with water contained in a thin tube. A **meniscus** is a concave water surface due to the attraction of the water molecules adjacent to the glass walls. No meniscus is present with mercury because of its weak attraction to the walls. Mercury in a glass tube has a convex surface due to surface tension.

As we have noted before, water, because of hydrogen bonding, has some very unusual properties. It will dissolve a great number of substances, both ionic and polar covalent. This is because of its polarity and ability to form hydrogen bonds. It has a high **heat capacity** (see Chapter 6). This is the heat absorbed to cause a specific increase in temperature. Water has both a high **heat of vaporization** and a high heat of fusion. These are the energies necessary to transform a liquid into a gas and the energy necessary to convert a solid to a liquid. Both thermal properties are a result of the strong hydrogen bonding between the molecules. Water has a high surface tension for the same reason. Finally, the fact that the solid form of water (ice) is less dense than liquid water is because of hydrogen bonds. These forces hold the water molecules in a rigid open crystalline framework. As ice starts to melt, the crystal structure breaks, and water molecules fill the holes in the structure. Filling the holes increases the density. The density reaches a maximum at around 4°C, and then the increasing kinetic energy of the particles causes the density to decrease.

Solids

At the macroscopic level, a **solid** is a substance that has both a definite volume and a definite shape. At the microscopic level, solids may be one of two types: amorphous or crystalline. **Amorphous solids** lack extensive ordering of the particles. There is a lack of regularity of the structure. There may be small regions of order separated by large areas of disordered particles. They resemble liquids more than solids in this characteristic. Amorphous solids have no distinct melting point because they are already liquids. They simply become softer and softer as the temperature rises. Amorphous solids are really liquids with exceptionally high viscosity. Glass, some plastics, and charcoal are examples of amorphous solids.

 BTW
In looking at crystal lattice diagrams, count all the particles in all three dimensions that surround another particle.

Crystalline solids display a very regular ordering of the particles in a three-dimensional structure called the **crystal lattice**. In this crystal lattice there are repeating units called **unit cells**. See the internet for diagrams of unit cells.

There are five types of crystalline solids:

BTW
Particles not in the center of a cell are shared by more than one cell. The contribution from these particles will be only a fraction of the complete particle.

- **Atomic solids** have individual atoms held in place by London forces. The noble gases are the only atomic solids known. Since the group is so small, atomic solids are sometimes grouped with molecular solids.

- **Molecular solids** have their lattices composed of molecules held in place by London forces, dipole-dipole forces, and hydrogen bonding. Solid methane and water (ice) are examples of molecular solids.

- **Ionic solids** have their lattices composed of ions held together by the attraction of opposite charges of the ions. These crystalline solids tend to be strong with high melting points due to the strength of the intermolecular forces. NaCl and other salts are examples of ionic solids.

- **Metallic solids** have metal atoms occupying the crystal lattice held together by metallic bonding. In **metallic bonding**, the electrons of the atoms are delocalized and free to move throughout the entire solid. This explains the electrical and thermal conductivity as well as many of the other properties of metals.

- **Network covalent solids** have covalent bonds joining the atoms together in the crystal lattice, which is quite large. Graphite, diamond, and silicon dioxide, SiO_2, are examples of network solids. There are fewer examples in this category than in the other categories except the atomic solids.

Phase Changes

An equilibrium exists between a liquid and its vapor. This is just one of several equilibria that exist between the states of matter. A **phase diagram** is a graph representing the relationship of all the states of matter of a substance. One type of phase diagram relates the states of a single substance to temperature and pressure. This type allows us to predict which state of matter will exist at a certain temperature and pressure combination. The following figure shows a general form of a phase diagram.

Note that the diagram has three general areas corresponding to the three states of matter: solid, liquid, and gas. The line from A to C and beyond represents the change in vapor pressure of the solid with temperature for the **sublimation** (going directly from a solid to a gas without first becoming a liquid) equilibrium. The A to D line and beyond represents the variation in the melting point with pressure. The A to B line represents the variation of the vapor pressure of a liquid with temperature and pressure. The terminal B point shown on this phase diagram is the **critical point** of the substance, the point beyond which the gas and liquid phases are indistinguishable from each other. At or beyond this critical point, no matter how much pressure is applied, it is not possible to condense the gas into a liquid. Point A is the

triple point of the substance, the combination of temperature and pressure at which all three states of matter can exist in equilibrium.

A General Phase Diagram

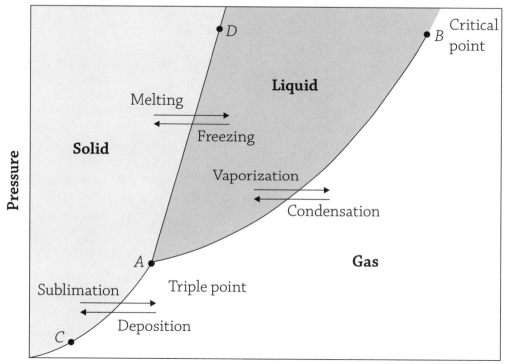

Intermolecular forces can affect phase changes. Strong intermolecular forces require more kinetic energy to convert a liquid into a gas. Stronger intermolecular forces make it easier to condense a gas into a liquid.

EASY MISTAKE

In looking at phase diagrams, be careful when moving from point to point to pay attention to any phase changes that might occur.

▶ Let's analyze a typical problem. Determine the strongest type of intermolecular force present in each of the following:

> methane, CH_4
>
> methyl alcohol, CH_3OH
>
> diamond, C
>
> methyl fluoride, CH_3F
>
> iron, Fe
>
> ammonium fluoride, NH_4F
>
> krypton difluoride, KrF_2
>
> sodium chloride, NaCl

▶ You may or may not encounter a question with eight substances, such as this question, but regardless of the number of substances, the procedure will be the same.

▶ For each of the substances the possible answers are ionic bonding, covalent bonding, metallic bonding, hydrogen bonding, dipole-dipole force, or London force. Forces, such as ion-dipole forces and ion-induced dipole forces, are not choices because these require the presence of two or more substances. For example, sodium chloride cannot use either of these two forces, but sodium chloride dissolved in water can. (Sodium chloride in water exhibits ion-dipole forces.)

▶ Each of the eight substances will exhibit London forces since they are present in everything containing electrons. London forces are only the strongest type of intermolecular force if there are no other attractions present. The most convenient method of analyzing this problem is to leave consideration of London forces to the last.

▶ We can begin with any of the intermolecular forces other than London forces. It is usually easiest to begin with the "normal" bonds (covalent, ionic, and metallic). Bonds only occur in specific circumstances. For example, metallic bonds only occur in metals or metal alloys. The only metal or alloy in the eight-substance list is iron. For this reason, the strongest intermolecular force in iron is metallic bonding.

▶ Very few materials use covalent bonding as an important intermolecular force. The best-known examples are silicon dioxide, SiO_2, graphite, C, and diamond, C. One of these three common examples, diamond, is in our question. Therefore, the strongest intermolecular force in diamond is covalent bonding.

▶ Ionic bonding is present in compounds containing a metal and a nonmetal or in a compound containing one or more polyatomic ions. There are a few exceptions to this generalization, but these usually do not appear in this type of question. An ionic substance must contain at least two different elements, so we know that iron and diamond cannot involve ionic bonding. This is a check on the validity of our earlier predictions. Several of the remaining substances, methane, methyl alcohol, methyl fluoride, and krypton difluoride, do not contain a metal or a polyatomic ion. (Be careful with methyl alcohol. This compound does not contain the hydroxide ion. The –OH group in this compound is an alcohol group.) Only two substances remain: ammonium fluoride and sodium chloride. Ammonium fluoride does not contain a metal, but it does contain a polyatomic ion, the ammonium ion. The presence of the polyatomic ion means that the strongest intermolecular force will be ionic bonding. Sodium chloride consists of a metal and a nonmetal, which means that its strongest intermolecular force is ionic bonding.

▶ The next strongest type of intermolecular force is hydrogen bonding. This type of force requires a hydrogen atom bonded to nitrogen, oxygen, or fluorine. We can eliminate any compound not containing hydrogen, KrF_2, and any compound not containing nitrogen, oxygen, or fluorine, CH_4. This leaves us with methyl alcohol, CH_3OH, and methyl fluoride, CH_3F. (You may need to sketch the atom arrangement for these two compounds to make the prediction.) In methyl alcohol, there is a bond between one of the hydrogen atoms and the oxygen atom. This attachment allows the formation of hydrogen bonds. In this compound, the remaining hydrogen atoms bond to the carbon. In methyl fluoride, the hydrogen atoms bond to the carbon. Since no hydrogen atoms are bonded to the fluorine, hydrogen bonding is not possible.

▶ We now have three substances remaining: methane, CH_4, methyl fluoride, CH_3F, and krypton difluoride, KrF_2. We also have two types of intermolecular force remaining: dipole-dipole forces and London forces. In order to match these substances and forces we **must know** which of the substances are polar and which are nonpolar. Polar substances use dipole-dipole forces, while nonpolar substances use London forces. To determine the polarity of each substance, we must draw a Lewis structure for the substance (Chapter 9) and use valence-shell electron pair repulsion (VSEPR) (Chapter 10). The Lewis structures for these substances follow:

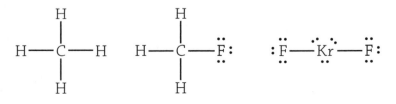

▶ The molecular geometries of methane and methyl fluoride are tetrahedral. In the case of methane, this symmetrical arrangement of polar covalent carbon-hydrogen bonds leads to a canceling of the bond polarities, resulting in a nonpolar molecule. As a nonpolar molecule, the strongest intermolecular force in methane is a London force. In methyl fluoride, a fluorine atom replaces one of the hydrogen atoms in methane. The polarity of the polar covalent carbon-fluorine bond is not equal to that of a carbon-hydrogen bond. This difference in polarity means that the bond polarities will no longer cancel. If the bond polarities do not cancel, the molecule is polar. Methyl fluoride is a polar molecule. The strongest intermolecular force in the polar methyl fluoride is a dipole-dipole force.

▶ The krypton atom in krypton difluoride does not obey the octet rule. The presence of five pairs around the krypton leads to a trigonal bipyramidal electron-group geometry. The presence of three lone pairs and two bonding pairs around the krypton makes the molecule linear. The two krypton-fluorine bonds are polar covalent. However, in a linear molecule, the bond polarities pull directly against each other and cancel. Cancelled bond polarities make the molecule nonpolar. The strongest intermolecular force in the nonpolar krypton difluoride is London force.

▶ The answers are:

methane, CH_4,	London force
methyl alcohol, CH_3OH	hydrogen bonding
diamond, C	covalent bonding
methyl fluoride, CH_3F	dipole-dipole force
iron, Fe	metallic bonding
ammonium fluoride, NH_4F	ionic bonding
krypton difluoride, KrF_2	London force
sodium chloride, NaCl	ionic bonding

EXERCISES

EXERCISE 11-1

Answer the following questions.

1. Which of the following is not an example of a change of state?
    a. melting
    b. freezing
    c. burning
    d. boiling
    e. subliming

2. Ion-dipole forces exist in which of the following?
    a. $NaCl(s)$
    b. $KNO_3(s)$
    c. $CH_3OH(l)$
    d. $NaCl(aq)$
    e. $HCl(g)$

3. Dipole-dipole forces exist in all of the following *except*:
    a. $HBr(g)$
    b. $KBr(s)$
    c. $H_2S(g)$
    d. $CHCl_3(l)$
    e. $H_2O(l)$

4. Hydrogen bonding exists in all of the following *except*:
    a. $CH_3F(g)$
    b. $H_2O(g)$
    c. $NH_3(g)$
    d. $HF(g)$
    e. $CH_3CH_2OH(g)$

5. In which of the following are London dispersion forces the most important intermolecular force present?
    a. $H_2O(l)$
    b. $NH_3(l)$
    c. $HBr(l)$
    d. $NaCl(l)$
    e. $CH_4(l)$

6. The thin "skin" over the surface of a liquid is due to the _____.

7. Viscosity is the _____.

8. Capillary action is the _____.

9. Water has a high heat of vaporization because of _____.

10. Amorphous solids lack _____.

11. Crystalline solids display _____.

12. Which type of solid is ice?
    a. atomic
    b. molecular
    c. ionic
    d. metallic
    e. network covalent

13. Which type of solid is diamond?
    a. atomic
    b. molecular
    c. ionic
    d. metallic
    e. network covalent

14. Which type of solid is solid xenon?

     a.   atomic

     b.   molecular

     c.   ionic

     d.   metallic

     e.   network covalent

15. Which type of solid is sodium chloride?

     a.   atomic

     b.   molecular

     c.   ionic

     d.   metallic

     e.   network covalent

16. Which type of solid is sodium?

     a.   atomic

     b.   molecular

     c.   ionic

     d.   metallic

     e.   network covalent

17. Sketch a general phase diagram.

18. The two fixed points on a phase diagram are the _____ and the _____.

19. What is the strongest type of intermolecular force present in each of the following?

     a.   hydrogen chloride, $HCl(g)$

     b.   hydrazine, $NH_2NH_2(l)$

     c.   graphite, $C(s)$

     d.   calcium fluoride, $CaF_2(s)$

     e.   methane, $CH_4(g)$

20. What is the strongest type of intermolecular force present in each of the following?
    a. sulfur tetrafluoride, $SF_4(g)$
    b. ethyl alcohol, $CH_3CH_2OH(l)$
    c. methyl fluoride, $CH_3F(g)$
    d. ammonium phosphate, $(NH_4)_3PO_4(s)$
    e. xenon tetrafluoride, $XeF_4(g)$

Solutions

MUST KNOW

 A solution is a homogeneous mixture composed of a solvent and one or more solutes.

A solution is a homogeneous mixture composed of a solvent and one or more solutes.

The solubility of solids in liquids normally increases with increasing temperature, but the reverse is true of gases dissolving in liquids. The solubility of gases in liquids increases with increasing pressure.

Colligative properties are those properties of solutions that depend on the number of solute particles present and not their identity.

In this chapter we're going to learn how to work with the concepts associated with solutions: concentration units, solubility, and especially colligative properties. We will also examine the properties of colloids. If you are still unsure about calculations and the mole concept, review Chapters 1, 3, and 4. (We said the mole concept would be important!)

Concentration Units

A **solution** is a homogeneous mixture composed of solvent and one or more solutes. The **solvent** is normally the substance present in the greatest amount. The **solute** is normally the substance that is present in the smaller amount. If water is the solvent, it is an **aqueous solution**. You may have more than one solute in a solution.

Some substances will dissolve in a particular solvent and others will not. There is a general rule in chemistry that states *like dissolves like*. Polar substances (such as alcohols) will dissolve in polar solvents like water. Nonpolar solutes (such as iodine) will dissolve in nonpolar solvents such as carbon tetrachloride. The mass of solute per 100 mL of solvent (g/100 mL) is a common alternative to expressing the solubility as molarity. It is necessary to specify the temperature because the solubility of a substance will vary with the temperature. The solubility of a solid dissolving in a liquid normally increases with increasing temperature. The reverse is true for a gas dissolving in a liquid.

A solution containing the maximum amount of solute per given amount of solvent at a given temperature is a **saturated solution**. An **unsaturated solution** has less than that maximum amount of solute dissolved. Sometimes, there may be more than that maximum amount of solute, resulting in a **supersaturated solution**. Supersaturated solutions are unstable and eventually expel the excess solute, forming a saturated solution.

There are many ways of expressing the relative amounts of solute(s) and solvent in a solution. The terms *saturated*, *unsaturated*, and *supersaturated* discussed previously give a qualitative measure of solubility, as do the terms *dilute* and *concentrated*. **Dilute** refers to a solution that has a relatively small amount of solute in comparison to the amount of solvent. **Concentrated** refers to a solution that has a relatively large amount of solute in comparison to the solvent. These terms are very subjective, and chemists prefer to use quantitative ways of expressing the concentration of solutions. A number of these concentration units prove to be useful, including percentage, molarity, molality, and mole fraction.

Percentage

One common way of expressing the relative amount of solute and solvent is through percentage, amount per hundred. There are three ways that we may express this percentage: mass percent, mass/volume percent, and volume/volume percent.

- **mass (weight) percentage**   The mass percentage of a solution is the mass of the solute divided by the mass of the solution and then multiplied by 100 to yield percentage.

  mass% = (grams of solute/grams solution) × 100%

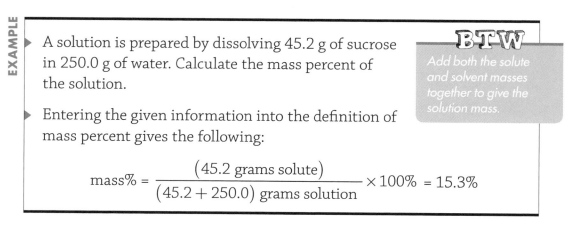

EXAMPLE

A solution is prepared by dissolving 45.2 g of sucrose in 250.0 g of water. Calculate the mass percent of the solution.

▸ Entering the given information into the definition of mass percent gives the following:

$$\text{mass\%} = \frac{(45.2 \text{ grams solute})}{(45.2 + 250.0) \text{ grams solution}} \times 100\% = 15.3\%$$

BTW

Add both the solute and solvent masses together to give the solution mass.

- **mass/volume percentage** The mass/volume percent of a solution is the mass of the solute (typically in grams) divided by the volume (typically in mL) of the solution and then multiplied by 100 to yield percentage.

  mass/volume% = (mass solute/volume of solution) × 100%

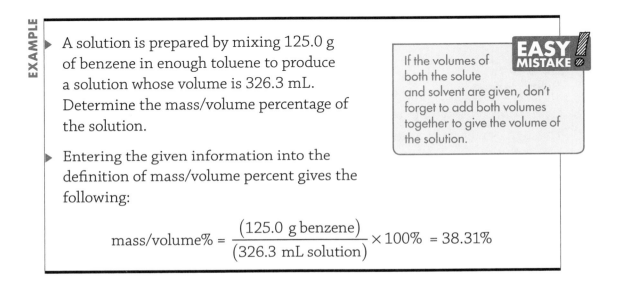

EXAMPLE

A solution is prepared by mixing 125.0 g of benzene in enough toluene to produce a solution whose volume is 326.3 mL. Determine the mass/volume percentage of the solution.

EASY MISTAKE

If the volumes of both the solute and solvent are given, don't forget to add both volumes together to give the volume of the solution.

Entering the given information into the definition of mass/volume percent gives the following:

$$\text{mass/volume\%} = \frac{(125.0 \text{ g benzene})}{(326.3 \text{ mL solution})} \times 100\% = 38.31\%$$

- **volume/volume percentage** The third case is one in which both the solute and solvent are liquids. The volume percent of the solution is the volume of the solute divided by total volume of the solution and then again multiplied by 100 to generate the percentage.

  volume% = (volume solute/volume solution) × 100%

EXAMPLE

▶ Determine the volume percentage of carbon tetrachloride in a solution prepared by dissolving 100.0 mL of carbon tetrachloride and 100.0 mL of methylene chloride in 750.0 mL of chloroform. Assume the volumes are additive.

▶ Entering the given information into the definition of volume percent gives the following:

$$\text{volume\%} = \frac{(100.0 \text{ mL carbon tetrachloride})}{(100.0 + 100.0 + 750.0) \text{ mL solution}} \times 100\% = 10.53\%$$

Molarity

While percentage concentration is common in everyday life, most chemists (and chemistry students) use molarity. **Molarity (M)** is the number of moles of solute per liter of solution:

M = mol solute/L solution

Refer back to Chapter 4 for problems related to molarity.

BTW

In preparing a molar solution, the correct number of moles of solute (commonly converted to grams using the molar mass) is dissolved and diluted to the required volume.

Molality

Molality (m) is the moles of solute per kilogram of solvent.

m = mol solute/kg solvent

Notice that it is kilograms of solvent, not solution. In the other concentration units, it has been the mass or volume of the entire solution. In molal solutions, it is the mass of the solvent only.

EXAMPLE

▶ Ethylene glycol ($C_2H_6O_2$) is in antifreeze. Determine the molality of ethylene glycol in a solution prepared by adding 31.0 g of ethylene glycol to 200.0 g of water.

It is the kilograms of the solvent in the denominator and not the solution.

▶ Entering the given information and two conversions into the definition of molality gives the following:

$$\text{molality} = \left(\frac{31.0 \text{ g } C_2H_6O_2}{200.0 \text{ g } H_2O} \right)\left(\frac{1000 \text{ g}}{1 \text{ kg}} \right)\left(\frac{1 \text{ mol } C_2H_6O_2}{62.1 \text{ g } C_2H_6O_2} \right)$$
$$= 2.50 \text{ } m \text{ } C_2H_6O_2$$

Mole Fraction

The mole fraction, X, is the moles of solute divided by the total number of moles of solution. Mole fraction was introduced in Chapter 5.

X = moles solute/total moles present

EXAMPLE

▶ Suppose you have a solution containing 0.50 mol of sodium chloride in 2.0 mol of water. What is the mole fraction of sodium chloride?

BTW

The sum of all the mole fractions of the components of a solution must equal 1.00.

▶ Entering the given information into the definition of mole fraction gives the following:

$$X_{NaCl} = \text{mol NaCl}/ (\text{mol NaCl} + \text{mol } H_2O)$$
$$= 0.50/(0.50 + 2.00) = 0.20$$

Dilution

Another way to prepare a solution is by diluting a more concentrated solution to a more dilute one by adding solvent. You can use the following equation:

$$(C_{before}) (V_{before}) = (C_{after}) (V_{after})$$

In this equation, C represents any concentration unit and V is any volume unit, if the same concentration and volume unit are used on both sides.

EXAMPLE

▶ Determine the final concentration when 800.0 mL of water is added to 300.0 mL of a 0.1000 M solution of HCl. Assume the volumes are additive.

▶ Label each of the given values, then enter the given information into the rearranged dilution equation:

M_{before} = 0.1000 M V_{before} = 300.0 mL

M_{after} = ? V_{after} = (300.0 + 800.0) mL = 1100.0 mL

M_{after} = $(M_{before}) (V_{before})/(V_{after})$
     = (0.1000 M) (300.0 mL)/(1100.0 mL)
     = 0.02727 M

Temperature and Pressure Effects on Solubility

The solubility of most solids increases with increasing temperature. However, the solubility of gases in liquids decreases with increasing temperature. For example, if you open a cold bottle of soda and a warm bottle of soda, more gas is released by the warm soda. This is the basis of thermal pollution, in which the solubility of oxygen in stream or lake water is decreased if the water is polluted by heat.

The solubility of liquids and solids are, in general, not affected by changes in pressure. However, the solubility of gases is affected. The greater the partial pressure of the gas, the greater the solubility of the gas in a liquid.

Colligative Properties

Some of the properties of solutions are dependent on the chemical and physical nature of the individual solute. However, there are solution properties that depend only on the *number* of solute particles and not their identity. These properties are **colligative properties** and they include:

- vapor pressure lowering

- freezing point depression

- boiling point elevation

- osmotic pressure

Vapor Pressure Lowering

If a liquid is placed into a sealed container, molecules will evaporate from the surface of the liquid and will eventually establish a gas phase over the liquid that is in equilibrium with the liquid phase. The partial pressure of this gas is the **vapor pressure** of the liquid. This vapor pressure is temperature dependent: the higher the temperature, the higher the vapor pressure. If a solution is prepared, then the solvent contribution to the vapor pressure of the solution depends on the vapor pressure of the pure solvent, $P°_{solvent}$, and the mole fraction of the solvent. We can find the contribution of solvent to the vapor pressure of the solution by the following relationship:

$$P_{solvent} = X_{solvent} \, P°_{solvent}$$

A similar calculation gives the solute contribution.

$$P_{solute} = X_{solute} \, P°_{solute}$$

There may be more than one solute present. If there is more than one solute, we find the contribution of each solute in the same way. If the solute is nonvolatile, $P°_{solute} = 0$.

The vapor pressure of a solution is the sum of the contributions of all solutes and the solvent.

$$P_{solution} = P_{solvent} + P_{solute} \quad \text{or} \quad P_{solution} = X_{solvent} \, P°_{solvent} + X_{solute} \, P°_{solute}$$

This relationship is **Raoult's law**.

Not all solutions obey Raoult's law. Any solution that follows Raoult's law is an **ideal solution**. However, many solutions are not ideal solutions. A solution may have a vapor pressure higher than predicted by Raoult's law or a solution may have a vapor pressure lower than predicted by Raoult's law. Solutions with a lower than expected vapor pressure are showing a negative deviation from Raoult's law, while those with a higher than expected vapor pressure are showing a positive deviation from Raoult's law.

In general, ideal solutions result when the intermolecular forces between the particles are like those in the solvent or solute alone. When the intermolecular forces in the solution are weaker, the molecules tend to escape more readily and produce a positive deviation. If the intermolecular forces in the solution are greater than those in the individual constituents, then the particles stay together instead of vaporizing. These solutions show a negative deviation.

If a pure liquid is the solvent and you add a nonvolatile solute ($P°_{solute} = 0$), the vapor pressure of the resulting solution is always less than the pure liquid. The addition of the solute lowers the vapor pressure and the amount of lowering is proportional to the number of solute particles added.

There is an even distribution of solvent particles throughout the solution, even at the surface. There are fewer solvent particles at the gas-liquid interface. Evaporation takes place at this interface. Fewer solvent particles escape into the gas phase and thus the vapor pressure is lower. The higher the concentration of solute particles, the less solvent is at the interface and the lower the vapor pressure.

Freezing Point Depression and Boiling Point Elevation

The freezing point of a solution of a nonvolatile solute is always lower than the pure solvent, and the boiling point is always higher. It is the number of solute particles that determines the amount of the lowering of the freezing point and raising of the boiling point.

The amount of lowering of the freezing point is proportional to the molality of the solute and is given by the equation:

$$\Delta T_f = i\, K_f \times m$$

ΔT_f is the number of degrees that the freezing point has been lowered (the difference in the freezing point of the pure solvent and the solution). **K_f** is the freezing point depression constant (a constant of the individual solvent). The **molality (m)** is the molality of the solute, and **i** is the van't Hoff factor, which is the ratio of the number of moles of particles released into solution per mole of solute dissolved. For a nonelectrolyte such as sucrose, the van't Hoff factor would be 1. For an electrolyte such as sodium sulfate, you must take into consideration that if one mole of Na_2SO_4 dissolves, three moles of particles would result (2 mol Na^+, 1 mol SO_4^{2-}). Therefore, the van't Hoff factor should be 3. However, because sometimes there is a pairing of ions in solution the observed van't Hoff factor is slightly less. The more dilute the solution, the closer the observed van't Hoff factor should be to the expected value.

Just as the freezing point of a solution is always lower than the pure solvent, the boiling point of a solution is always higher than the solvent. The relationship is like the earlier one for the freezing point depression:

$$\Delta T_b = i\,K_b \times m$$

In this equation, ΔT_b is the number of degrees that the boiling point has been elevated (the difference between the boiling point of the pure solvent and the solution), K_b is the boiling point elevation constant, m is the molality of the solute, and i is again the van't Hoff factor.

Osmotic Pressure

A U-tube contains a solution and pure solvent. A **semipermeable membrane** separates the two components. Such a membrane allows the passage of solvent molecules but not solute particles. This arrangement will result in the level of the solvent side decreasing while the solution side is increasing. This indicates that the solvent molecules are passing through the

The freezing point depression, ΔT_f, is subtracted from the normal freezing point of the solvent to determine the actual freezing point of the solution. The boiling point elevation, ΔT_b, is added to the normal boiling point of the solvent to determine the actual boiling point of the solution.

semipermeable membrane. This process is **osmosis**. Eventually the system would reach equilibrium and the difference in levels would remain constant. The difference in the two levels is related to the **osmotic pressure**. In fact, one could exert a pressure on the solution side exceeding the osmotic pressure. This will cause the solvent molecules to move back through the semipermeable membrane into the solvent side. This process is **reverse osmosis** and is the basis of desalination of seawater for drinking purposes.

The osmotic pressure is a colligative property and mathematically represented as:

$$\pi = (nRT/V)i$$

In this equation, π is the osmotic pressure in atmospheres, n is the number of moles of solute, R is the ideal gas constant (0.0821 L·atm/K·mol), T is the Kelvin temperature, V is the volume of the solution, and i is the van't Hoff factor. If one knows the moles of solute and the volume in liters, n/V may be replaced by the molarity, M. It is possible to calculate the molar mass of a solute from osmotic pressure measurements. This is especially useful in the determination of the molar mass of large molecules such as proteins.

Colloids

Particles will settle out of water from a muddy stream. This combination is a heterogeneous mixture, where the particles are large (in excess of 10^3 nm in diameter) and is a **suspension**. On the other hand, dissolving sodium chloride in water produces a true homogeneous solution, where the solute particles are less than 1 nm in diameter. Particles do not settle out of a true solution, because of their very small particle size. However, there is a mixture with particle diameters falling between solutions and suspensions. These are **colloids** and have particles in the 1 to 10^3 nm diameter range. Smoke, fog, milk, mayonnaise, and latex paint are all examples of colloids.

Many times, it is difficult to distinguish a colloid from a true solution. The most common way is by shining a light through the mixture. A light shone through a true solution is invisible, but a light shown through a colloid is visible due to the reflection of the light off the larger colloid particles. This is the **Tyndall effect**.

Colligative Properties Problems

Let's examine a few examples of how to approach colligative property problems. We will begin with a Raoult's law example.

EXAMPLE

▶ What is the vapor pressure of a solution made by mixing 80.0 g of chloroform, $CHCl_3$, in 800.0 g of carbon tetrachloride, CCl_4? The vapor pressure of chloroform is 197 torr, and the vapor pressure of carbon tetrachloride is 114 torr (all vapor pressures are determined at 25°C).

▶ Raoult's law requires the mole fraction of each volatile material. Whenever the mole fraction is not present, we will need to calculate the value from the number of moles. In this example, we must begin by calculating the moles of each constituent of the solution.

$$\text{Moles chloroform} = (80.0 \text{ g } CHCl_3)(1 \text{ mol } CHCl_3/119.378 \text{ g } CHCl_3) = 0.6701402 \text{ mol } CHCl_3 \text{ (unrounded)}$$

▶ Moles carbon tetrachloride = $(800.0 \text{ g } CCl_4)(1 \text{ mol } CCl_4/153.823 \text{ g } CCl_4) = 5.2007827 \text{ mol } CCl_4$ (unrounded)

▶ The mole fraction of chloroform (solute) is:

$$X_{solute} = \frac{(0.6701402) \text{ mol } CHCl_3}{(0.6701402 + 5.2007827) \text{ mol}} = 0.1141456 \text{ (unrounded)}$$

▶ It is possible to calculate the mole fraction of carbon tetrachloride (solvent) in a similar manner. However, simply subtracting the mole fraction of chloroform from 1 will give the same value.

$$X_{solvent} = 1 - 0.1141456 = 0.8858544$$

▶ Using Raoult's law:

$$P_{solution} = X_{solvent} P^{\circ}_{solvent} + X_{solute} P^{\circ}_{solute}$$

$$P_{solution} = (0.8858544)(114 \text{ torr}) + (0.1141456)(197 \text{ torr})$$
$$= 123.47408 = 123 \text{ torr}$$

In our next example, we will show an example of freezing point depression and boiling point elevation. This will require us to use the equation $\Delta T_f = i\, K_f m$. In this example, we will use a nonelectrolyte, so we will not need the van't Hoff factor (or simply $i = 1$).

> Determine both the freezing point and boiling point of a solution containing 15.50 g of naphthalene, $C_{10}H_8$, in 0.200 kg of benzene, C_6H_6. Pure benzene freezes at 278.65 K and boils at 353.25 K. K_f for benzene is 5.07 K/m and its K_b is 2.64 K/m.

> We will begin by calculating the change in the freezing point, ΔT_f, to answer this problem. The problem gives us the value of K_f, and we are assuming that $i = 1$. Therefore, we need to know the molality of the solution to find our answer. To determine the molality, we will begin by determining the moles of naphthalene. Naphthalene has a molar mass of 128.17 g/mol. Thus, the number of moles of naphthalene present is:

$$\frac{15.50 \text{ g}}{128.17 \ \dfrac{\text{g}}{\text{mol}}} = 0.120933 \text{ mol (unrounded)}$$

> The molality of the solution, based on the definition of molality, would be:

$$\frac{0.120933 \text{ mol}}{0.200 \text{ kg}} = 0.604665 \ m \text{ (unrounded)}$$

> We can enter the given values along with the calculated molality into the freezing point depression equation:

You must subtract the ΔT value from the normal freezing point to get the freezing point of the solution.

$$\Delta T = i\, K_f m = (1)\ (5.07 \ K/m)\ (0.604665 \ m)$$
$$= 3.06565 \ K = 3.07 \ K$$

$$T_f = (278.65 - 3.07) \ K = 275.58 \ K \text{ (or } 2.43°C)$$

> To calculate the boiling point of the solution, we use the relationship:

$$\Delta T_b = i\, K_b \times \text{molality}$$

> We already have the van't Hoff factor, the K_b, and solution molality so we can simply substitute:

$$\Delta T_b = i\, K_b \times \text{molality} = (1)\,(2.64\ K/m)\,(0.604665\ m)$$
$$= 1.59632\ K\ (\text{unrounded})$$

$$T_b = (353.25 + 1.59632)\ K = 354.85\ K\ (\text{or}\ 81.70°C)$$

The freezing point depression and boiling point elevation techniques are useful in calculating the molar mass of a solute or its van't Hoff factor. In these cases, you will begin with the answer (the freezing point depression or the boiling point elevation) and follow the same steps as previously in reverse order.

In the next example, we will examine the colligative property of osmotic pressure. This will require us to use the relationship $\pi = i(nRT/V)$.

EXAMPLE

> A solution prepared by dissolving 7.95 mg of a gene fragment in 25.0 mL of water has an osmotic pressure of 0.295 torr at 25.0°C. Assuming the fragment is a nonelectrolyte, determine the molar mass of the gene fragment.

> In this example, we need to determine the molar mass (g/mol) of the gene fragment. This requires two pieces of information: the mass of the substance and the number of moles. We know the mass (7.95 mg), and we need to determine the number of moles present. We will rearrange the osmotic pressure relationship to $n = \pi\, V/RT$. We know the solute is a nonelectrolyte, so $i = 1$. We can now enter the given values into the rearranged equation and perform a pressure and a volume conversion:

$$\left[\frac{(0.295 \text{ torr})\ (25.0 \text{ mL})}{\left(0.0821\ \dfrac{\text{L} \cdot \text{atm}}{\text{mol} \cdot \text{K}}\right)(298.2 \text{ K})}\right]\left(\frac{1 \text{ atm}}{760 \text{ torr}}\right)\left(\frac{1 \text{ L}}{1000 \text{ mL}}\right)$$

$$= 3.97 \times 10^{-7} \text{ mol}$$

▶ The final step in the problem is to combine the given mass and the moles we found to give the molar mass. This will require the conversion of the milligrams given into grams.

$$\left[\frac{7.95}{3.97 \times 10^{-7} \text{ mol}}\right]\left(\frac{10^{-3} \text{ g}}{1 \text{ mg}}\right) = 2.00 \times 10^4 \text{ g/mol}$$

In the preceding examples, we saw how to deal with nonelectrolytes. If the solution contains an electrolyte, there will only be one change necessary. This change will be to enter the value of the van't Hoff factor. We will see how to do this in the next example.

EASY MISTAKE

A molar mass (molecular weight) less than 1.0 g/mol indicates an error. No molecular weight is less than the atomic weight of hydrogen.

EXAMPLE

▶ Determine both the freezing point and boiling point of a solution containing 15.50 g of sodium sulfate, Na_2SO_4, in 0.200 kg of water. Pure water freezes at 0.00°C and boils at 100.00°C. K_f for water is 1.86°C/m and its K_b is 0.52°C/m.

▶ We will follow the same procedure as in the earlier naphthalene/ benzene example. You may wish to look over these examples in parallel to see exactly where the difference between an electrolyte and nonelectrolyte manifests itself. We will again begin by calculating the freezing point, ΔT_f. The problem gives us the value of K_f. In solution, the strong electrolyte, sodium sulfate, ionizes as:

$$Na_2SO_4(aq) \rightarrow 2\ Na^+(aq) + SO_4^{2-}(aq)$$

▶ From this relationship, we can see that each Na_2SO_4 produces three ions. The production of three ions means that the van't Hoff factor, i, is 3. We need to know the molality of the solution to find our answer. To determine the molality, we will begin by determining the moles of sodium sulfate. Sodium sulfate has a molar mass of 142.04 g/mol. Thus, the number of moles of sodium sulfate present is:

$$\frac{15.50 \text{ g}}{142.04 \ \frac{\text{g}}{\text{mol}}} = 0.10912419 \text{ mol (unrounded)}$$

▶ The molality of the solution, based on the definition of molality, would be:

$$\frac{0.10912419 \text{ mol}}{0.200 \text{ kg}} = 0.545620951 \text{ mol (unrounded)}$$

▶ We can enter the given values along with the calculated molality into the freezing point depression equation:

$$\Delta T_f = i\,K_f m = (3)\,(1.86°C/m)\,(0.545620951 \ m)$$
$$= 3.0445649°C = 3.04°C$$

$$T_f = (0.00 - 3.04)°C = -3.04°C$$

▶ To calculate the boiling point of the solution we use the relationship:

$$\Delta T_b = i\,K_b \times m$$

▶ We already have the van't Hoff factor, K_b, and solution molality so we can simply substitute:

$$\Delta T_b = i\,K_b \times m = (3)\,(0.52°C/m)$$
$$(0.545621 \ m) = 0.85116868°C$$
$$\text{(unrounded)}$$

$$T_b = (100.00 + 0.85116868)°C = 100.85°C$$

EASY MISTAKE

The most common error made in colligative property problems is to forget to separate the ions of an electrolyte. The van't Hoff factor, even when not needed, is a useful reminder.

EXERCISES

EXERCISE 12-1

Answer these questions about solutions.

1. What phrase applies to a situation where a polar solvent dissolves a polar solute?

2. A solution containing the maximum amount of solute per given amount of solvent at a given temperature is said to be:
   a. concentrated
   b. dilute
   c. unsaturated
   d. saturated
   e. supersaturated

3. All methods of numerically expressing the concentration of a solution contain a term in the denominator referring to the entire solution *except*:
   a. mass percentage
   b. molarity
   c. molality
   d. mole fraction
   e. volume percentage

4. During a dilution, the one factor that remains constant is the:
   a. quantity of solute
   b. quantity of solvent
   c. volume
   d. density
   e. osmotic pressure

5. Raoult's law, osmotic pressure, and freezing point depression calculations use, without conversion, which of the following respective concentration units?
   a. molarity, molality, and mole fraction
   b. mole fraction, molarity, and molality
   c. mole fraction, molality, and molarity
   d. molarity, mole fraction, and molality
   e. molality, molarity, and mole fraction

6. The simplest way to distinguish a colloid from a solution is the:
   a. Tyndall effect
   b. osmotic pressure
   c. density
   d. vapor pressure
   e. freezing point

7. A solution is prepared by dissolving 10.0 g of table salt (sodium chloride) and 15.0 g of cane sugar (sucrose) in 500.0 g of water. Calculate the mass percent of the table salt in the solution.

8. Determine the molality of ethanol, C_2H_5OH, in a solution prepared by adding 50.0 g of ethanol to 250.0 g of water.

9. What is the vapor pressure of a solution made of chloroform, $CHCl_3$, in carbon tetrachloride, CCl_4? The mole fraction of chloroform is 0.250. At 25°C the vapor pressure of chloroform is 197 torr, and the vapor pressure of carbon tetrachloride is 114 torr.

10. Determine both the freezing point and boiling point of a 1.50 m aqueous solution of ammonium phosphate, $(NH_4)_3PO_4$. Pure water freezes at 0.00°C and boils at 100.00°C. K_f for water is 1.86°C/m and its K_b is 0.52°C/m.

11. A solution prepared by dissolving 6.95×10^{-3} g of protein in 0.0300 L of water has an osmotic pressure of 0.195 torr at 25°C. Assuming the protein is a nonelectrolyte, determine the molar mass of the gene fragment.

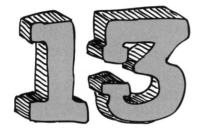  **Kinetics**

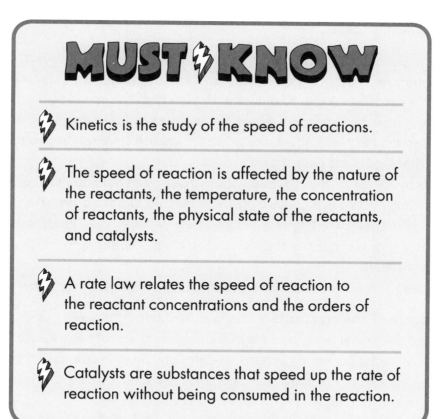

MUST ⚡ KNOW

⚡ Kinetics is the study of the speed of reactions.

⚡ The speed of reaction is affected by the nature of the reactants, the temperature, the concentration of reactants, the physical state of the reactants, and catalysts.

⚡ A rate law relates the speed of reaction to the reactant concentrations and the orders of reaction.

⚡ Catalysts are substances that speed up the rate of reaction without being consumed in the reaction.

We are now going to learn about kinetics—those factors that affect the speed of reactions. We will be discussing the concept of half-lives; you will see this concept again in Chapter 20 on nuclear chemistry. It will be necessary in some of the problems to solve for an exponential quantity along with the use of the *ln* and the e^x functions, so you might want to refresh yourself with your calculator manual. And don't forget—chemistry is not a spectator sport!

Reaction Rates

Many times, we can use thermodynamics to predict whether a reaction will occur spontaneously, but it gives very little information about the speed at which a reaction occurs. **Kinetics** is the study of the speed of reactions. It is largely an experimental science. Some general qualitative ideas about reaction speed may be applied, but accurate quantitative relationships require that we collect experimental data.

For a chemical reaction to occur, there must be a collision between the reactants at the correct place on the molecule, the **reactive site**. That collision is necessary to transfer kinetic energy to break old chemical bonds and reform new ones. If the collision doesn't transfer enough energy, no reaction will occur.

In general, five factors can affect the rates of a chemical reaction:

- **nature of the reactants**   Large, slow-moving complex molecules will tend to react slower than smaller ones because there is a greater chance of collisions occurring somewhere else on the molecule rather than the reactive site. Also, if the molecules are slow moving, the number of collisions will be smaller.

- **the temperature**   Increasing the temperature normally increases the reaction rate since each species has a higher kinetic energy and the number of collisions is increased. This increases the chance that enough energy will be transferred during collisions to cause the reaction.

- **the concentration of reactants**   Increasing the concentration of the reactants (or pressure, if gases are involved) normally increases the reaction rate due to the increased number of collisions.

- **physical state of reactants**   Gases and liquids tend to react faster than solids because of the increase in surface area of the gases and liquids versus the solid.

- **catalysts**   Using a **catalyst** increases the reaction rate.

 IRL There are many, many biological catalysts that allow reactions to occur in living organisms without extreme conditions of heat, pressure, and so on. These biological catalysts are called enzymes.

The rate of reaction is the change in concentration per change in time. It is possible to find the rate of reaction from a graph of concentration of a reactant versus time. The procedure involves drawing a tangent to the curve at the point in the reaction where we wish to know the rate.

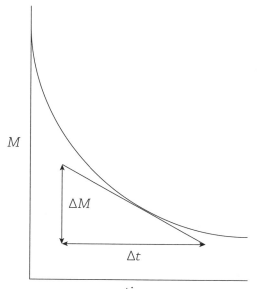

The slope of the tangent ($\Delta M/\Delta t$) is the instantaneous rate of the reaction at this time. To determine the rate at a different time, we would need to draw another tangent line. In most kinetic studies, we wish to know the initial rate. The initial rate comes from a tangent drawn to the curve at the very beginning of the reaction.

Rate Laws

Let's consider those cases in which the reactant concentration may affect the speed of reaction. For the general reaction:

$$a\,A + b\,B + \ldots \rightarrow c\,C + d\,D + \ldots$$

Where the lowercase letters are the coefficients in the balanced chemical equation, the uppercase letters stand for the reactant, and product chemical species and initial rates are used, then the rate equation (rate law) is as follows:

$$\text{rate} = k[A]^m[B]^n \ldots.$$

In this expression, k is the **rate constant** (for the chemical reaction at a given temperature). The exponents, m and n (<u>not</u> a and b from the chemical equation), are the **orders of reaction**. The orders indicate what effect a change in concentration of that reactant species will have on the reaction rate. If, for example, $m = 1$ and $n = 2$, then if the concentration of reactant A is doubled, the rate will also double ($[2]^1 = 2$), and if the concentration of reactant B is doubled, the rate will increase fourfold ($[2]^2 = 4$). This reaction is first order with respect to reactant A and second order with respect to reactant B. In this example, the rate equation would then be:

$$\text{rate} = k[A]^1[B]^2 = k[A][B]^2$$
(If the exponent is 1, it is generally not shown.)

Many times, we calculate the overall order of reaction: it is simply the sum of the individual coefficients, third order in this example. If the concentration of a reactant changes and that has no effect on the rate of reaction, then the reaction is zero order with respect to that reactant ($[2]^0 = 1$).

Once the rate has been determined, the orders of reaction can be determined by conducting a series of reactions in which we change the concentrations of the reactant species one at a time. We then mathematically determine the effect on the reaction rate. Once the orders of reaction have been determined, we calculate the rate constant.

> **BTW**
>
> *The rate law (the rate, the rate constant, and the orders of reaction) is determined experimentally.*

EXAMPLE

▶ Consider the reaction:

$$2\,NO(g) + O_2(g) \rightarrow 2\,NO_2(g)$$

▶ We collected the following kinetics data:

Experiment	Initial [NO]	Initial [O$_2$]	Rate of NO$_2$ formation (M/s)
1	0.01	0.01	0.05
2	0.02	0.01	0.20
3	0.01	0.02	0.10

▶ There are a couple of ways that you might interpret the preceding data to determine the rate equation. If the numbers involved are simple, then one can reason out the orders of reaction by varying the concentration of one reactant at a time. You can see that in going from experiment 1 to experiment 2, the [NO] doubles ([O$_2$] held constant) and the rate increases fourfold. This means that the reaction is second order with respect to NO. Comparing experiments 1 and 3, you see that the [O$_2$] doubles ([NO] was held constant) and the rate doubled. Therefore, the reaction is first order with respect to O$_2$ and the rate equation is:

$$rate = k[NO]^2\,[O_2]$$

▶ The rate constant can be determined by substituting the values of the concentrations of NO and O_2 from any of the experiments into the preceding rate equation and solving for k.

▶ Using experiment 1:

$$0.05 \ M/s = k \ (0.01 \ M)^2(0.01 \ M)$$

$$k = 0.05 \ M/s/(0.01 \ M)^2(0.01 \ M)$$

$$k = 5 \times 10^4/M^2s$$

▶ You can calculate a k for each experiment and average the values if needed.

▶ However, sometimes because of the complexity of the numbers, you must manipulate the equations mathematically. We use the ratio of the rate expressions of two experiments to determine the reaction orders. We choose the equations so that the concentration of only one reactant has changed while the others remain constant. In the preceding example, we will use the ratio of experiments 1 and 2 to determine the effect of a change of the concentration of NO on the rate. Then we will use experiments 1 and 3 to determine the effect of O_2. We cannot use experiments 2 and 3, since both chemical species have changed concentration.

▶ Comparing experiments 1 and 2:

$$\frac{rate_1 = k \ [NO]_1^m \ [O_2]_1^n}{rate_2 = k \ [NO]_2^m \ [O_2]_2^n} = \frac{0.05 \ \frac{M}{s} = k \ [0.01]^m \ [0.01]^n}{0.20 \ \frac{M}{s} = k \ [0.02]^m \ [0.01]^n}$$

▶ Canceling the rate constants and the $[0.01]^n$ and simplifying gives:

$$\frac{1}{4} = \left(\frac{1}{2}\right)^m$$

▶ Thus, $m = 2$, which you can check (you can use logarithms to solve for m). Comparing experiments 1 and 3:

$$\frac{0.05 \; \dfrac{M}{s} = k\,[0.01]^m\,[0.01]^n}{0.10 \; \dfrac{M}{s} = k\,[0.01]^m\,[0.02]^n}$$

▶ Canceling the rate constants and the $[0.01]^m$ and simplifying gives:

$$\frac{1}{2} = \left(\frac{1}{2}\right)^n \quad \text{Thus, } n = 1.$$

▶ Therefore, the rate equation is rate = $k[NO]^2[O_2]$. The rate constant, k, could be determined by choosing any of the three experiments, substituting the concentrations, rate, and orders into the rate expression, and then solving for k.

Integrated Rate Laws (Time and Concentration)

So far, we have used only instantaneous data in the rate expression. These expressions allow us to answer questions concerning the speed of the reaction at a specific moment, but not questions about how long it might take to use up a certain reactant. However, if we consider changes in the concentration of reactants or products over time, as expressed in the **integrated rate laws**, we can answer these types of questions.

Consider the following reaction: A ⟶ B. Assuming this reaction is first order, we can express the rate of reaction as the change in concentration of reactant A with time:

$$\text{rate} = -\frac{\Delta[A]}{\Delta t} \text{ in addition to the rate law: Rate} = k[A]$$

Setting these terms equal gives:

$$-\frac{\Delta[A]}{\Delta t} = k\,[A] \text{ and integrating over time gives:}$$

$$\ln\frac{[A]_0}{[A]_t} = kt$$

In this equation, ln is the natural logarithm, $[A]_0$ is the initial concentration of a reactant, $[A]_t$ is the concentration of the same reactant at some time t, k is the rate constant, and t is the time passed between $[A]_0$ and $[A]_t$.

If the reaction is second order in A, we can derive the following equation using the same procedure:

$$\frac{1}{[A]_t} - \frac{1}{[A]_0} = kt$$

The terms have the same meaning as in the first-order equation.

EXAMPLE

▶ Hydrogen iodide, HI, decomposes through a second-order process to the elements. The rate constant is $2.40 \times 10^{-21}/M$ s at 25°C. How long will it take for the concentration of HI to drop from 0.200 M to 0.190 M at 25°C?

▶ This is a simple plug-in problem, with $k = 2.40 \times 10^{-21}/M$ s, $[A]_0 = 0.200$ M, and $[A]_t = 0.190$ M. You may simply plug in the values and then solve for t, or you first rearrange the equation to give $t = (1/[A]_t - 1/[A]_0)/k$. You will get the same answer, 1.10×10^{20} s, in either case. If you get a negative answer, you incorrectly interchanged $[A]_t$ and $[A]_0$.

The order of reaction can be determined graphically by using the integrated rate law. If a plot of the ln[A] versus time yields a straight line, then the reaction is first order with respect to reactant A. If a plot of 1/[A] versus time yields a straight line, then the reaction is second order with respect to reactant A.

The reaction **half-life, $t_{1/2}$**, is the amount of time that it takes for a reactant concentration to decrease to one-half its initial concentration. For a first-order reaction, the half-life is a constant, independent of reactant concentration, and has the following relationship:

$$t_{1/2} = \frac{0.693}{k}$$

For second-order reactions, the half-life does depend on the initial reactant concentration. We calculate it using the following formula:

$$t_{1/2} = \frac{1}{k\,[A]_0}$$

Radioactive decay is a first-order process. See Chapter 20 for a discussion of half-lives related to nuclear reactions and other information on radioactivity.

EXAMPLE

▶ The rate constant for the radioactive decay of thorium-232 is 5.0 × 10^{-11}/year. Determine the half-life of thorium-232.

▶ This radioactive decay process follows first-order kinetics. Substitute the value of k into the appropriate equation:

$$t_{1/2} = \frac{0.693}{k} = \frac{0.693}{5.0 \times 10^{-11}\ \text{yr}^{-1}} = 1.4 \times 10^{10}\ \text{yr}$$

Arrhenius and Activation Energy

A change in the reaction temperature affects the rate constant k. As the temperature increases, the value of the rate constant increases and the reaction is faster. The Swedish scientist Arrhenius derived a relationship that related the rate constant and temperature. The Arrhenius equation has the form: $k = A\, e^{-E_a/RT}$. In this equation, k is the rate constant and A is a term called the frequency factor that accounts for molecular orientation. The symbol e is the natural logarithm base and R is universal gas constant (in energy terms). Finally, T is the Kelvin temperature and E_a is the **activation energy**, the minimum amount of energy needed to initiate or start a chemical reaction. This process is shown in the following figure. Also notice that since the products are of a lower energy than the reactants, this is an exothermic reaction.

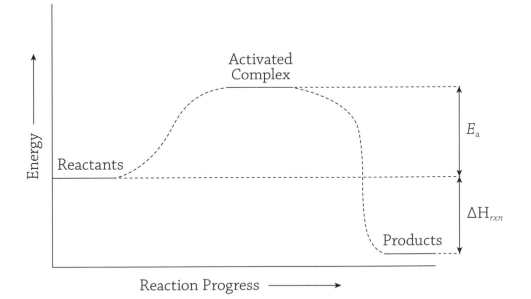

We commonly use the Arrhenius equation to calculate the activation energy of a reaction. One way to do this is to plot the ln of k versus $1/T$. This gives a straight line whose slope is E_a/R. Knowing the value of R, we can calculate the value of E_a.

Another method for determining the activation energy involves using a modification of the Arrhenius equation. If we try to use the Arrhenius equation directly, we have one equation with two unknowns (the frequency factor and the activation energy). The rate constant and the temperature are experimental values, while R is a constant. One way to avoid this difficulty is to perform the experiment twice. We determine experimental values of the rate constant at two different temperatures. We then assume that the frequency factor is the same at these two temperatures. We now have a new equation derived from the Arrhenius equation that allows us to calculate the activation energy. This equation is:

$$\ln\frac{k_1}{k_2} = \frac{E_a}{R}\left[\frac{1}{T_1} - \frac{1}{T_2}\right]$$

The two rate constant values, k_1 and k_2, are the values determined at two different temperatures, T_1 and T_2. The temperatures must be in Kelvin. The units on the rate constants will cancel, leaving a unitless ratio. R is 8.314 J/mol·K. The activation energy will have units of joules/mol.

BTW

Remember that the k in the denominator, k_2, goes with the first temperature, T_2.

EXAMPLE

▶ The variation in the rate constant at two different temperatures for the decomposition of HI(g) to $H_2(g) + I_2(g)$ is given as follows. Calculate the activation energy.

EASY MISTAKE

Calculations containing logarithms and/or differences in reciprocals are very sensitive to rounding. Be careful to not round intermediate values.

T (K)	k (L/mol·s)
555	3.52×10^{-7}
781	3.95×10^{-2}

▶ We will use the equation:

$$\ln\frac{k_1}{k_2} = \frac{E_a}{R}\left[\frac{1}{T_1} - \frac{1}{T_2}\right]$$

$T_1 = 781$ K

$k_1 = 3.95 \times 10^{-2}$ L/mol·s

$T_2 = 555$ K

$k_2 = 3.52 \times 10^{-7}$ L/mol·s

$R = 8.314$ J/mol·K

$E_a = ?$

▶ We could rearrange this equation and then enter the values, or we can enter the values and then rearrange the equation. Using the second method gives:

$$\ln\frac{3.95 \times 10^{-2} \text{ L / mol} \cdot \text{s}}{3.52 \times 10^{-7} \text{ L / mol} \cdot \text{s}} = \frac{E_a}{8.314 \text{ J / mol} \cdot \text{K}}\left[\frac{1}{555 \text{ K}} - \frac{1}{781 \text{ K}}\right]$$

$\ln(1.122159 \times 10^5) =$

$$\frac{E_a}{8.314 \text{ J / mol} \cdot \text{K}}\left[\left(1.80180 \times 10^{-3} - 1.2804097 \times 10^{-3}\right)\left(\frac{1}{K}\right)\right]$$

$$11.62818 = \frac{E_a}{8.314 \text{ J / mol}}\left[5.21392 \times 10^{-4}\right]$$

$$E_a = \frac{(11.62818)(8.314 \text{ J / mol})}{(5.21392 \times 10^{-4})} = 1.85420 \times 10^5$$

$$= 1.9 \times 10^5 \text{ J/mol}$$

High activation energies are normally associated with slow reactions. Anything done to lower the activation energy of a reaction will tend to speed up the reaction.

Catalysis

A **catalyst** is a substance that speeds up the rate of reaction without being consumed in the reaction. A catalyst may take part in the reaction and even change during the reaction, but at the end of the reaction, it is at least theoretically recoverable in its original form. It will not produce more of the product, but it allows the reaction to proceed more quickly. In equilibrium reactions, the catalyst speeds up both the forward and reverse reactions. Catalysts speed up the rates of reaction by lowering the activation energy of the reaction. In general, there are two distinct types of catalysts.

Homogeneous catalysts are catalysts that are in the same phase or state of matter as the reactants. They provide an alternate reaction pathway with a lower activation energy.

The decomposition of hydrogen peroxide is a slow, one-step reaction, especially if the solution is kept cool and in a dark bottle:

$$2\ H_2O_2 \rightarrow 2\ H_2O + O_2$$

However, adding ferric ion speeds up the reaction tremendously.

$$2\ Fe^{3+} + H_2O_2 \rightarrow 2\ Fe^{2+} + O_2 + 2\ H^+$$
$$2\ Fe^{2+} + H_2O_2 + 2\ H^+ \rightarrow 2\ Fe^{3+} + 2\ H_2O$$

Notice that in the reaction, the catalyst, Fe^{3+}, reduced to Fe^{2+} in the first step of the mechanism. In the second step, oxidation of Fe^{2+} back to Fe^{3+} occurred. Overall then, the catalyst remains unchanged. Notice also that although the catalyzed reaction is a two-step reaction, it is significantly faster than the original uncatalyzed one-step reaction.

A **heterogeneous catalyst** is in a different phase or state of matter than the reactants. Most commonly, the catalyst is a solid and the reactants are liquids or gases. These catalysts provide a surface for the reaction. The reactant on the surface is more reactive than the free molecule. Many times, these heterogeneous catalysts are finely divided metals.

IRL Chemists use an iron catalyst in the Haber process, which converts nitrogen and hydrogen gases into ammonia. The automobile catalytic converter is another example. A solid platinum oxide catalyst helps convert oxides of nitrogen to nitrogen and oxygen, which are not as harmful as the original nitrogen oxides.

Mechanisms

Many reactions proceed from reactants to products through a sequence of reactions. This sequence of reactions is the **reaction mechanism**. For example, consider the reaction:

$$A + 2B \rightarrow E + F$$

Most likely, E and F do not form from the simple collision of an A and two B molecules. This reaction might follow this reaction sequence:

$$A + B \rightarrow C$$
$$C + B \rightarrow D$$
$$D \rightarrow E + F$$

If you add together the preceding three equations, you will get the overall equation $A + 2B \rightarrow E + F$. C and D are **reaction intermediates**, chemical species that are produced and consumed during the reaction, but that do not appear in the overall reaction.

   Each individual reaction in the mechanism is an **elementary step** or **reaction**. Each of these reaction steps has its own individual rate of reaction. One of the reaction steps will be slower than the others and is the **rate-determining step**. This rate-determining step limits how fast the overall reaction can occur. Therefore, the rate law of the rate-determining step is the rate law of the overall reaction.

It is possible to determine the rate equation for an elementary step directly from the stoichiometry. This will not work for the overall reaction. The reactant coefficients in an elementary step become the reaction orders in the rate equation for that elementary step.

Many times, a study of the kinetics of a reaction gives clues to the reaction mechanism. For example, consider the following reaction:

$$NO_2(g) + CO(g) \rightarrow NO(g) + CO_2(g)$$

(It has been determined experimentally that the rate law for this reaction is rate = $k[NO_2]^2$.)

The rate law indicates that the reaction does not occur with a simple collision between NO_2 and CO. The reaction might follow this mechanism:

$$NO_2(g) + NO_2(g) \rightarrow NO_3(g) + NO(g)$$
$$NO_3(g) + CO(g) \rightarrow NO_2(g) + CO_2(g)$$

Notice that if you add these two steps together, you get the overall reaction. We have determined that the first step is the slow step in the mechanism, the rate-determining step. If we write the rate law for this elementary step it is: rate = $k[NO_2]^2$, identical to the experimentally determined rate law for the overall reaction.

Note that both steps in the mechanism are **bimolecular reactions**, reactions that involve the collision of two chemical species. **Unimolecular reactions** are reactions in which a single chemical species decomposes or rearranges (radioactive decay is an example). Both bimolecular and unimolecular reactions are common, but the collision of three or more chemical species (termolecular) is quite rare. Thus, in developing or assessing a mechanism, it is best to consider only unimolecular or bimolecular elementary steps.

More About Rate Law and Half-Life ($t_{1/2}$)

To work problems of this type it is important to know what the order of the reaction is. This may come directly from the problem; for example, "the reaction is first order." In some cases, the order comes indirectly from the problem, for example, "the rate law is rate = $k[A]^2$," which means the reaction is second order.

In most problems, it will be to your advantage to find and label as many of the following terms as possible: $[A]_0$, $[A]_t$, k, t, and $t_{1/2}$. You will not need all these terms for all problems. However, the ones you have, or do not have, often direct you toward the solution.

Let's apply the preceding two paragraphs to an example problem.

EXAMPLE

▶ The first-order decomposition of gaseous dinitrogen pentoxide, N_2O_5, to nitrogen dioxide, NO_2, and oxygen, O_2, has a rate constant of 4.9×10^{-4} s^{-1} at a certain temperature. Calculate the half-life of this reaction.

▶ This is a first-order reaction. The following relationships apply to first-order reactions:

$$\text{rate} = k[A] \qquad \ln\frac{[A]_0}{[A]_t} = kt \qquad t_{1/2} = \frac{0.693}{k}$$

▶ These are the only equations we have exclusive to first-order reactions. For this reason, one or more of these equations will be necessary to work the problem. (In this problem, A = N_2O_5.)

▶ Using the list of terms ($[A]_0$, $[A]_t$, k, t, and $t_{1/2}$), and the given information, we find that $k = 4.9 \times 10^{-4}$ s^{-1} and $t_{1/2} = ?$ (We are using a "?" to indicate the term we are seeking.) The only first-order relationship we have that relates k and $t_{1/2}$ is $t_{1/2} = \dfrac{0.693}{k}$.

This significantly limits our options to finishing the problem. To finish the problem, we need to enter k into the relationship:

$$t_{1/2} = \frac{0.693}{k} = \frac{0.693}{4.9 \times 10^{-4} \ s^{-1}} = 1414.2857 = 1.4 \times 10^3 \ s$$

Instead of finding the half-life, we can find the rate constant for a reaction.

EXAMPLE

▶ A substance undergoes a simple decomposition. The reaction is first order. In one experiment, at 25°C, the concentration of this substance decreased from 1.000 M to 0.355 M after 4.25 min. What was the rate constant for this reaction?

▶ In this case, as in the preceding example, we have a first-order reaction. This limits us to the same set of three relationships as in the earlier example. Using the list of terms ($[A]_0$, $[A]_t$, k, t, and $t_{1/2}$), and the given information, we find k = ?, $[A]_0$ = 1.000 M, $[A]_t$ = 0.355, and t = 4.25 min. (We also know T = 25°, but since T is not one of the five terms, it is irrelevant.) The only first-order relationship with these four terms is $\ln\dfrac{[A]_0}{[A]_t}$ = kt. We can either enter the given values into this equation, or we can rearrange the equation before entering the values.

▶ The latter method is usually preferable, and we shall employ it in this example. We rearrange the equation to find the rate constant, k, and enter the appropriate values:

$$k = \frac{\ln\dfrac{[A]_0}{[A]_t}}{t} = \frac{\ln\dfrac{[1.000]_0}{[0.355]_t}}{4.25 \text{ min}} = \frac{1.035637}{4.25 \text{ min}} = 0.243679$$

$$= 0.244 \text{ min}^{-1}$$

We will now examine a problem that is not first order.

▶ Molecules of butadiene (C_4H_6) will dimerize to C_8H_{12}. The rate law is second order in butadiene, and the rate constant is 0.014 L/mol·s. How many seconds will it take for the concentration of butadiene to drop from 0.010 M to 0.0010 M?

▶ This is a second-order reaction. The following relationships apply to order reactions:

$$\text{rate} = k[A]^2 \qquad \frac{1}{[A]_t} - \frac{1}{[A]_0} = kt \qquad t_{1/2} = \frac{1}{k[A]_0}$$

▶ These are the only equations we have exclusive to second-order reactions. For this reason, one or more of these equations will be necessary to work the problem. (In this problem, A is butadiene.)

▶ Using the list of terms ($[A]_0$, $[A]_t$, k, t, and $t_{1/2}$) and the given information, we find that k = 0.014 L/mol · s, $[A]_0$ = 0.010 M, $[A]_t$ = 0.0010 M, and t = ? sec. The only second-order relationship with these four terms is: $\frac{1}{[A]_t} - \frac{1}{[A]_0} = kt$. We can rearrange this relationship to isolate the time and enter the given values:

$$t = \frac{\dfrac{1}{[A]_t} - \dfrac{1}{[A]_0}}{k} = \frac{\dfrac{1}{[0.0010]_t} - \dfrac{1}{[0.010]_0}}{0.014 \text{ L / mol} \cdot \text{s}}$$

$$= \frac{(1000 - 100)\text{L / mol}}{0.014 \text{ L / mol} \cdot \text{s}} = \frac{(900)}{0.014 \text{ s}^{-1}} = 64285.7 = 6.4 \times 10^4 \text{ s}$$

EXERCISES

EXERCISE 13-1

Answer these questions about kinetics.

1. What five factors can affect the rate of a chemical reaction?

2. The rate of a reaction increases by a factor of nine when the concentration of one of the reactants is tripled. The order with respect to this reactant is _____.

3. The rate of a reaction does not change when the concentration of one of the reactants is tripled. The order with respect to this reactant is _____.

4. What is the equation for the half-life of a second-order reaction?

5. A student calculates the activation energy for a reaction and gets −0.12 J/mol. What is wrong with this answer?

6. True or False: An automobile catalytic converter is an example of a heterogeneous catalyst.

7. True or False: You can get a rate law directly from a balanced chemical equation.

8. True or False: You can get a rate law directly from the rate-determining (slow) step in a mechanism.

9. True or False: Termolecular (three molecules) steps are common in mechanisms.

10. True or False: The mechanism for a reaction with the rate law, rate = $k[A]^2[B]$, will have a step where two molecules of A collide with a molecule of B.

11. A certain substance undergoes a first-order reaction with a rate constant of 1.25 hr^{-1}. How long will it take for the concentration of this reactant to drop from 1.0 M to 0.38 M?

12. Determine the half-life of a second-order reaction if the rate constant is 1.78 $M^{-1}s^{-1}$. The initial concentration is 0.575 M.

13. Nitrosyl chloride, NOCl, decomposes as follows:

$$2\ NOCl(g) \rightarrow 2\ NO(g) + Cl_2(g)$$

This reaction has a rate constant of $9.3 \times 10^{-5}\ M^{-1}\ s^{-1}$ at 100°C, and $1.0 \times 10^{-3}\ M\ s^{-1}$ at 130°C. Determine the value of the activation energy in kJ/mol.

Flashcard App

14 Chemical Equilibria

MUST KNOW

- A chemical equilibrium results when two exactly opposite reactions occur at the same place, at the same time, and with the same rate.

- An equilibrium constant expression represents the equilibrium system.

- Le Châtelier's principle describes the shifting of the equilibrium system due to changes in concentration, pressure, and temperature.

In this chapter we are going to take on the concept of chemical equilibria, the mathematical representations that we use in equilibrium systems and the manipulation of equilibrium by factors such as temperature and pressure. Chapters 15 and 16 will rely on the basic concepts presented in this chapter. Mastering them here will make things much easier later. This will require work on your part, though, so get ready!

Equilibrium

Very few chemical reactions proceed to completion, totally using up one or more of the reactants and then stopping. Most reactions behave in a different way. Consider the general reaction:

$$a A + b B \rightarrow c C + d D$$

Reactants A and B are forming C and D. The reaction proceeds until appreciable amounts of C and D form. As C and D form, it is possible for C and D to start to react to form A and B:

$$c C + d D \rightarrow a A + b B$$

These two reactions proceed until the two rates of reaction become equal. That is, the speed of production of C and D in the first reaction is equal to the speed of production of A and B in the second reaction. Since these two reactions are occurring simultaneously in the same container, the amounts of A, B, C, and D become constant.

A chemical equilibrium results when two exactly opposite reactions are occurring at the same place, at the same time, and with the same rates of reaction. When a system reaches the equilibrium state, the reactions do not stop. A and B are still reacting to form C and D; C and D are still reacting to form A and B. But because the reactions proceed at the same rate the amounts of each chemical species are constant. This state is a dynamic equilibrium state to emphasize the fact that the reactions are still occurring—it is a dynamic, not a static state. A double arrow instead of a single arrow indicates an equilibrium state. For the preceding reaction, it would be:

$$a\,A + b\,B \rightleftharpoons c\,C + d\,D$$

If the temperature is constant and the reaction is at equilibrium, then the ratio of the two reactions, the forward and reverse, should become a constant. This constant is the reaction quotient, Q, and has the following form:

$$Q_c = \frac{[C]^c\,[D]^d}{[A]^a\,[B]^b}$$

> **BTW**
> At equilibrium the concentrations of the chemical species are constant, but not necessarily equal. There may be a lot of C and D and a little A and B or vice versa. The concentrations are constant, unchanging, but not necessarily equal.

This reaction quotient is a fraction. The numerator is the product of the chemical species on the right-hand side of the equilibrium arrow, each one raised to the power of that species' coefficient in the balanced chemical equation. The denominator is the product of the chemical species on the left-hand side of the equilibrium arrow, each one raised to the power of that species' coefficient in the balanced chemical equation. It is called Q_c, in this case, since molar concentrations are being used. If this was a gas phase reaction, gas pressures could be used, and it would become a Q_p.

> **BTW**
> Remember: products over reactants and coefficients become exponents.

Equilibrium Constants (K)

We can write a reaction quotient at any point during the reaction, but the most meaningful point is when the reaction has reached equilibrium. At equilibrium, the reaction quotient becomes the equilibrium constant, K_c (or K_p if gas pressures are being used). We express this equilibrium constant simply as a number without units since it is a ratio of concentrations or pressures. In addition, the concentrations of solids, pure liquids, or solvents (not in solution) that appear in the equilibrium expression are assumed to be 1, since their concentrations do not change.

Consider the Haber process to produce ammonia: $N_2(g) + 3 H_2(g) \leftrightarrows 2 NH_3(g)$. The equilibrium constant expression is:

$$K_c = \frac{[NH_3]^2}{[N_2][H_2]^3}$$

If the partial pressures of the gases were used, then the K_p would be written in the following form:

$$K_p = \frac{P^2_{NH_3}}{P_{N_2} P^3_{H_2}}$$

There is a relationship between the K_c and the K_p: $K_p = K_c (RT)^{\Delta n_g}$ where R is the ideal gas constant and Δn_g is the change in the number of moles of gas in the reaction. In this equation, use R and T without units.

The numerical value of the equilibrium constant gives an indication of the extent of the reaction after reaching equilibrium. If K_c is large, then that means the numerator is much larger than the denominator and the reaction has produced a relatively large quantity of products (reaction lies far to the right). If K_c is small, then the numerator is much smaller than the denominator and not much product has been formed (reaction lies far to the left).

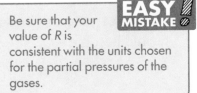

Be sure that your value of R is consistent with the units chosen for the partial pressures of the gases.

Le Châtelier's Principle

Henry Louis Le Châtelier discovered that if a chemical system at equilibrium is stressed (disturbed) it will reestablish equilibrium by shifting the rates of the reactions involved. This means that the amounts of the reactants and products will change, but the ratio will remain the same. One can stress the equilibrium in several ways: changes in concentration, pressure, and temperature. However, a catalyst will have no effect on the equilibrium amounts, since it affects both the forward and reverse reactions equally. It will simply allow the reaction to reach equilibrium faster.

> **IRL** Chemical engineers use Le Châtelier's principle to help increase the yield of product in chemical processes.

Changes in Concentration

If the stress to the equilibrium system is a change in concentration of one of the reactants or products, then the equilibrium will react to remove that stress. If, for example, we decrease the concentration of a chemical species, the equilibrium will shift to produce more of it. In doing so, the concentration of chemical species on the other side of the reaction arrows will be decreased. If the concentration of a chemical species is increased, the equilibrium will shift to consume it, increasing the concentration of chemical species on the other side of the reaction arrows.

For example, again consider the Haber process: $N_2(g) + 3\,H_2(g) \rightleftharpoons 2\,NH_3(g)$. If one increases the concentration of hydrogen gas, then the equilibrium shifts to the right in order to consume some of the added hydrogen. In doing so, the concentration of ammonia (NH_3) will increase and the concentration of nitrogen gas will decrease.

Changes in Pressure

Changes in pressure are only significant if there are gases involved. The pressure may be changed by changing the volume of the container or by changing the concentration of a gaseous species. If the container becomes smaller, the pressure increases because there

The concentrations may change, but the value of K_c or K_p would remain the same (unless the temperature changes).

are an increased number of collisions on the inside walls of the container. This stresses the equilibrium system, and it will shift in order to reduce the pressure. A shift toward the side of the equation that has the least number of moles of gas will accomplish this. If the container size is increased, the pressure decreases, and the equilibrium will shift to the side containing the greatest number of moles of gas in order to increase the pressure. If the number of moles of gas is the same on both sides, then changing the pressure will not influence the equilibrium.

Once again consider the Haber reaction: $N_2(g) + 3\,H_2(g) \leftrightarrows 2\,NH_3(g)$. Note that on the left side there are 4 moles of gas (1 of nitrogen and 3 of hydrogen) and 2 moles on the right. If the container is made smaller, the pressure increases, and the equilibrium will shift to the right because 4 moles of gas would be converted to 2 mol. In doing so, the concentrations of nitrogen and hydrogen gases would decrease, and the concentration of ammonia would increase.

Don't forget! Pressure effects are only important for gases.

Changes in Temperature

Changing the temperature really changes the amount of heat in the system and is like a concentration effect. In order to treat it in this fashion, one must know which reaction, forward or reverse, is exothermic (releasing heat).

Once again, let's consider the Haber reaction: $N_2(g) + 3\,H_2(g) \leftrightarrows 2\,NH_3(g)$. The formation of ammonia is exothermic (liberating heat), so that we could write the reaction as:

$$N_2(g) + 3\,H_2(g) \leftrightarrows 2\,NH_3(g) + \text{heat}$$

If the temperature of the reaction mixture is increased, the amount of heat increases, and the equilibrium would shift to the left in order to consume the added heat. In doing so, the concentration of nitrogen and hydrogen gases would increase, and the concentration of ammonia gas would decrease.

A change in the temperature is the only way to change the value of K.

EXAMPLE

▶ Given the following equilibrium (endothermic as written), predict what changes, if any, would occur if the following stresses are applied after equilibrium was established.

$$CaCO_3(s) \rightleftharpoons CaO(s) + CO_2(g)$$

▶ Add CO_2.   Left—the equilibrium shifts to remove some of the excess CO_2.

▶ Remove CO_2.   Right—the equilibrium shifts to replace some of the CO_2.

▶ Add CaO.   No change—solids do not shift equilibria unless they are totally removed.

▶ Increase T.   Right—endothermic reactions shift to the right when heated.

▶ Decrease V.   Left—a decrease in volume, or an increase in pressure, will shift the equilibrium toward the side with less gas.

▶ Add a catalyst.   No change—catalysts do not affect the position of an equilibrium.

Additional Equilibria Problems

Let's do a series of equilibria problems. You should pay close attention not only to what is different about each problem but also to what is the same.

Let's start off by determining an equilibrium constant from experimental data.

BTW

Equilibria problems require an equilibrium constant expression in nearly every case. You should begin each problem with this expression. This will get you started on the problem.

A container has the following equilibrium established at 700°C:

$$N_2(g) + 3\,H_2(g) \rightleftharpoons 2\,NH_3(g)$$

The equilibrium mixture had an ammonia concentration of 0.120 M, a nitrogen concentration of 1.03 M, and a hydrogen concentration of 1.62 M. Determine the value of the equilibrium constant, K_c.

> **BTW**
> Temperatures often appear in equilibria problems, but most problems will not require you to use the temperature.

The first step in this, and most equilibria problems, should begin by writing the equilibrium constant expression. In this case, the expression is:

$$K_c = \frac{\left[NH_3\right]^2}{\left[N_2\right]\left[H_2\right]^3}$$

The problem gives us $[NH_3]$ = 0.120 M, $[N_2]$ = 1.03 M, and $[H_2]$ = 1.62 M. The next step is to enter the given values into the equilibrium constant expression and enter the values into your calculator.

$$K_c = \frac{\left[NH_3\right]^2}{\left[N_2\right]\left[H_2\right]^3} = \frac{\left[0.120\right]^2}{\left[1.03\right]\left[1.62\right]^3} = 3.288367 \times 10^{-3}$$
$$= 3.29 \times 10^{-3}$$

Now let's solve for an equilibrium concentration given experimental data and the equilibrium constant.

The equilibrium value for the following equilibrium is 2.42×10^{-3} at a certain temperature.

$$N_2(g) + 3\,H_2(g) \rightleftharpoons 2\,NH_3(g)$$

Determine the ammonia concentration at equilibrium with 2.00 M nitrogen and 3.00 M hydrogen.

▶ The first step in this, and most equilibria problems, should begin by writing the equilibrium constant expression. In this case, the expression is:

$$K_c = \frac{[NH_3]^2}{[N_2][H_2]^3}$$

▶ The problem gives us $[N_2] = 2.00\ M$ and $[H_2] = 3.00\ M$. The next step is to enter the given values into the equilibrium constant expression and rearrange the expression to isolate the ammonia. Some people find it easier to reverse these steps.

$$K_c = \frac{[NH_3]^2}{[N_2][H_2]^3} = \frac{[NH_3]^2}{[2.00][3.00]^3} = 2.42 \times 10^{-3}$$

$$[NH_3]^2 = 2.42 \times 10^{-3}\ [2.00]\ [3.00]^3 = 0.13068\ \text{(unrounded)}$$

$$[NH_3] = \sqrt{0.13068} = 0.361497 = 0.361\ M\ NH_3$$

For gases, we can determine K_p (equilibrium constant using pressures) instead of a K_c.

EXAMPLE

▶ Determine the value of K_p for the following equilibrium: $C(s) + CO_2(g) \rightleftharpoons 2CO(g)$. At equilibrium, the carbon monoxide, CO, pressure is 1.22 atm and the carbon dioxide, CO_2, pressure is 0.780 atm.

▶ The first step in this, and most equilibria problems, should begin by writing the equilibrium constant expression. In this case, the expression is for a K_p so pressures must be present. The expression is:

$$K_p = \frac{P_{CO}^2}{P_{CO_2}}$$

▶ The problem gives us: P_{CO} = 1.22 atm, and P_{CO_2} = 0.780 atm. The next step is to enter the given values into the equilibrium constant expression and enter the values into your calculator.

$$K_p = \frac{P_{CO}^2}{P_{CO_2}} = \frac{(1.22)^2}{(0.780)} = 1.9082 = 1.91$$

In the preceding example, we calculated the K_p. Now, let's calculate the K_c.

▶ Calculate K_c for the preceding equilibrium at 25°C.

▶ This is one of the few problems where an equilibrium expression is not necessary. It is also one of the few cases where the temperature is important. We will begin this problem by writing the equation that relates K_p and K_c:

$$K_p = K_c \left(RT\right)^{\Delta n_g}$$

▶ We have the following values given to us: K_p = 1.91, T = 25°C, and R (from Chapter 5 on gas laws) = (0.0821 L·atm/mol·K). The presence of the Kelvin temperature unit in the value for R should serve as a reminder to convert °C to K. The temperature is 298 K.

BTW

The "g" subscript, which may not appear in this equation in your internet reference, is a reminder to focus only on the moles of gas present in the equilibrium reaction.

▶ We need to determine the value of Δn_g. We find this value from the reaction: $C(s) + CO_2(g) \leftrightarrows 2\ CO(g)$. We ignore the solid. There is 1 mol of gas on the reactant side and 2 mol of gas on the product side, therefore, Δn_g = (2 – 1).

▸ We can now enter these values into the equation to get:

$$1.91 = K_c \left[\left(0.0821 \ \frac{\text{L} \cdot \text{atm}}{\text{mol} \cdot \text{K}}\right)(298 \ \text{K})\right]^{(2-1)}$$

$$K_c = 0.07806816 = 0.0781$$

We will finish this section by examining one of the most common types of equilibrium problems.

EXAMPLE

▸ For the reaction $I_2(g) \rightleftharpoons 2 \ I(g)$ at a certain temperature, the value of K is 3.76×10^{-5}. If the initial I_2 concentration is $0.500 \ M$, what will be the equilibrium concentrations of $I_2(g)$ and $I(g)$?

▸ The first step in this, and most equilibria problems, should begin by writing the equilibrium constant expression. In this case, the expression is:

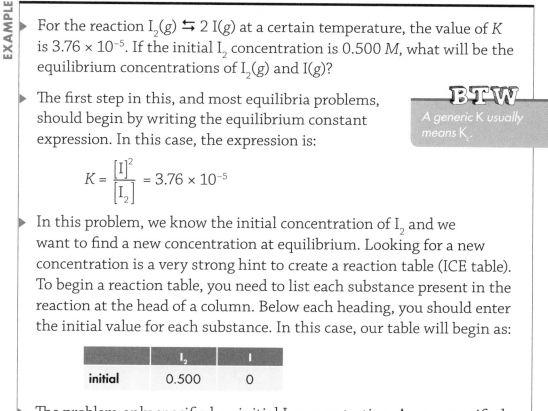

BTW

A generic K usually means K_c.

$$K = \frac{[I]^2}{[I_2]} = 3.76 \times 10^{-5}$$

▸ In this problem, we know the initial concentration of I_2 and we want to find a new concentration at equilibrium. Looking for a new concentration is a very strong hint to create a reaction table (ICE table). To begin a reaction table, you need to list each substance present in the reaction at the head of a column. Below each heading, you should enter the initial value for each substance. In this case, our table will begin as:

	I_2	I
initial	0.500	0

▸ The problem only specified an initial I_2 concentration. Any unspecified initial value appears as a zero in the table.

▶ The next line in the table will include information on how the initial values will change. A zero value in the table indicates how the change will occur. The equilibrium concentration of any substance can never be zero. Since the iodine atom concentration is initially equal to zero, we must add some quantity to this. The source of these iodine atoms must be the iodine molecules. This will result in a decrease in the concentration of iodine molecules. If we assume that the change in the molarity of iodine molecules is x, then based on the reaction stoichiometry, the iodine atom concentration change will be $2x$. This information begins the next line in our table:

	I_2	I
initial	0.500	0
change	$-x$	$+2x$

▶ To complete the table, we need to add each column to get the equilibrium values:

	I_2	I
initial	0.500	0
change	$-x$	$+2x$
equilibrium	$0.500 - x$	$+2x$

▶ This type of table is an initial/change/equilibrium (ICE) table. The acronym comes from the first letter of the names for each row of the table. The bottom row of the ICE table goes into your equilibrium expression:

$$K = \frac{[I]^2}{[I_2]} = \frac{[2x]^2}{[0.500 - x]} = 3.76 \times 10^{-5}$$

▶ There are two ways to complete the problem from this point. We will show you both methods.

▶ The first method involves an assumption. Anytime the value of the equilibrium constant is very large or very small, an assumption is possible. In this case, the constant is very small. A small equilibrium constant indicates that only a small number of iodine molecules will break apart to become iodine atoms. This means that x will be small. A small x value means that $0.500 - x$ will be close to $0.500\ M$. As an assumption, we will assume that $0.500 - x$ is $0.500\ M$.

▶ If we use our assumption in the equilibrium expression, it changes as follows:

$$K = \frac{[I]^2}{[I_2]} = \frac{[2x]^2}{[0.500 - x]} = \frac{[2x]^2}{[0.500]} = \frac{[4x^2]}{[0.500]} = 3.76 \times 10^{-5}$$

▶ We can now rearrange this equation and determine the value of x:

$$x = \sqrt{\frac{3.76 \times 10^{-5}\ (0.500)}{4}} = 2.1679 \times 10^{-3}\ \text{(unrounded)}$$

▶ To finish the problem, we must enter this value into the bottom line of our table:

	I_2	I
initial	0.500	0
change	$- x$	$+ 2x$
equilibrium	$0.500 - x$	$+ 2x$
equilibrium	$0.500 - 2.1679 \times 10^{-3}$	$+ 2\ (2.1679 \times 10^{-3})$

▶ The equilibrium concentration of $I_2 = 0.498\ M$ and the equilibrium concentration of I is $4.34 \times 10^{-3}\ M$. At this point you should determine the validity of the assumption. In this book, we will consider the assumption valid if no more than the last digit of the initial value is affected, which it does in this case.

EXTRA HELP

A+

If you do not or cannot use the assumption method, then it is necessary to do the problem the "long way." We will go back to this point.

BTW

If you enter your answers into the equilibrium expression, the result should be near K.

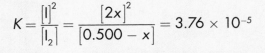

$$K = \frac{[I]^2}{[I_2]} = \frac{[2x]^2}{[0.500 - x]} = 3.76 \times 10^{-5}$$

EXAMPLE

▶ If we do not neglect the x in the denominator, we must rearrange this equation to:

$$4x^2 = (0.500 - x)(3.76 \times 10^{-5})$$

▶ We rearrange this to:

$$4x^2 + 3.76 \times 10^{-5}x - 1.88 \times 10^{-5} = 0$$

▶ This is in the quadratic form and, as such, requires you to use the quadratic equation.

$$x = \frac{-b \pm \sqrt{b^2 - 4ac}}{2a}$$

▶ In this case, $a = 4$, $b = 3.76 \times 10^{-5}$, and $c = -1.88 \times 10^{-5}$. If we enter these values into the quadratic equation, we get $x = 2.16 \times 10^{-3}$ or $x = -2.17 \times 10^{-3}$. The quadratic equation always gives two answers. We can eliminate one of the answers since it is physically impossible to have a negative concentration. This leaves us with only the positive root. If we enter this answer into the bottom line of our table, we get the following equilibrium concentrations:

$$[I_2] = 0.500 - 2.16 \times 10^{-3} = 0.498 \, M$$

$$[I] = 2(2.16 \times 10^{-3}) = 4.32 \times 10^{-3} \, M$$

EXERCISES

EXERCISE 14-1

Answer the following questions.

1. The values in a reaction quotient, Q, or an equilibrium constant expression, K, may be in terms of what two factors?

2. The terms in the numerator of an equilibrium constant expression are always the _____.

3. The reactants appear on which side of the reaction arrow in an equilibrium equation?

4. True or False: An equilibrium must respond to the stress created by a catalyst.

5. True or False: Increasing the pressure applied on the system is important for the following equilibrium.

$$H_2(g) + Br_2(g) \leftrightarrows 2\ HBr(g)$$

6. True or False: Adding calcium oxide, CaO, to the following equilibrium will have no effect.

$$CaCO_3(s) \leftrightarrows CaO(s) + CO_2(g)$$

7. True or False: Equilibrium constant expressions never include solids.

8. True or False: Water is important in equilibrium expressions for aqueous solutions.

9. Calculate K_p for the following equilibrium at 523 K:

$$NH_2COONH_4(s) \leftrightarrows 2\ NH_3(g) + CO_2(g)$$

At this temperature $K_c = 1.58 \times 10^{-8}$.

10. Nitrogen oxide, NO, reacts with hydrogen gas, H_2, to establish the following equilibrium:

$$2\ NO(g) + 2\ H_2(g) \leftrightarrows N_2(g) + 2\ H_2O(g)$$

The following concentrations of materials are sealed in a container: 0.250 M NO, 0.100 M H_2, and 0.200 M H_2O. Construct an ICE table for K_c.

11. The following equilibrium is established at 500 K:

$$S_2(g) + C(s) \leftrightarrows CS_2(g).$$

The initial pressure of sulfur vapor was 0.431 atm. Construct an ICE table for K_p.

12. Determine K_c for the following equilibrium:

$$2\ CH_4(g) \leftrightarrows C_2H_2(g) + 3\ H_2(g)$$

In order to determine this value, a mixture initially 0.0300 M in CH_4 was allowed to come to equilibrium. At equilibrium, the concentration of C_2H_2 was 0.01375 M.

13. The compound PH_3BCl_3 decomposes on heating by the following equilibrium:

$$PH_3BCl_3(s) \leftrightarrows PH_3(g) + BCl_3(g) \quad K_p = 1.60$$

Determine the equilibrium partial pressures of the gases resulting when a sample of PH_3BCl_3 is placed in a sealed flask and allowed to come to equilibrium.

14. Hydrogen sulfide gas decomposes at 700°C according to the following equilibrium:

$$2\ H_2S(g) \leftrightarrows 2\ H_2(g) + S_2(g)$$

At this temperature, K_c for this reaction is 9.1×10^{-8}, what will be the equilibrium concentrations of the gases if the initial concentration of H_2S was 0.200 M H_2S?

EXERCISE 14-2

Write the K_c expression for each of the following equilibria.

1. $CH_4(g) + 2\ O_2(g) \leftrightarrows CO_2(g) + 2\ H_2O(g)$

2. $Fe_2O_3(s) + 3\ CO(g) \leftrightarrows 2\ Fe(s) + 3\ CO_2(g)$

3. $HNO_2(aq) \leftrightarrows H^+(aq) + NO_2^-(aq)$

4. $Cu^{2+}(aq) + 4\ NH_3(aq) \leftrightarrows Cu(NH_3)_4^{2+}(aq)$

5. $ZnCO_3(s) \leftrightarrows Zn^{2+}(aq) + CO_3^{2-}(aq)$

EXERCISE 14-3

Write the K_p expression for each of the following equilibria.

1. $2\ NO_2(g) \leftrightarrows N_2(g) + 2\ O_2(g)$

2. $C(s) + H_2O(g) \leftrightarrows CO(g) + H_2(g)$

3. $2\ Zn(s) + O_2(g) \leftrightarrows 2\ ZnO(s)$

4. $2\ C_4H_{10}(g) + 13\ O_2(g) \leftrightarrows 8\ CO_2(g) + 10\ H_2O(l)$

5. $MnCO_3(s) \leftrightarrows CO_2(g) + MnO(s)$

EXERCISE 14-4

Indicate how the partial pressure of hydrogen will change if the following stresses are applied to the equilibrium below.

$$H_2(g) + CO_2(g) \rightleftharpoons H_2O(l) + CO(g)$$

$$\Delta H = +41 \text{ kJ}$$

1. Adding carbon dioxide

2. Adding water

3. Adding a catalyst

4. Increasing the temperature

5. Increasing the pressure

Flashcard App

15 Acids and Bases

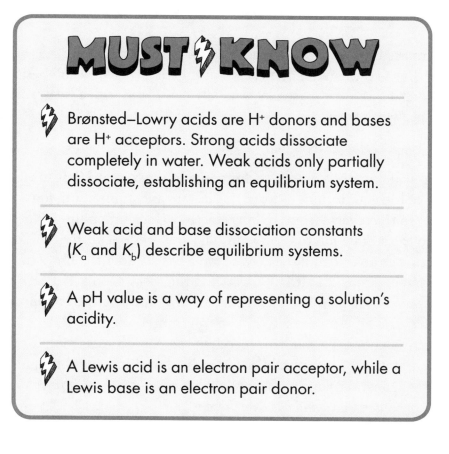

MUST ⚡ KNOW

⚡ Brønsted–Lowry acids are H^+ donors and bases are H^+ acceptors. Strong acids dissociate completely in water. Weak acids only partially dissociate, establishing an equilibrium system.

⚡ Weak acid and base dissociation constants (K_a and K_b) describe equilibrium systems.

⚡ A pH value is a way of representing a solution's acidity.

⚡ A Lewis acid is an electron pair acceptor, while a Lewis base is an electron pair donor.

n this chapter we're going to look at the equilibrium systems involving acids and bases. If you don't recall the Arrhenius acid-base theory, refer to Chapter 4 on aqueous solutions. We will learn a couple of other acid-base theories, the concept of pH, and will apply those basic equilibrium techniques we covered in Chapter 14 to acid-base systems. In addition, you will need to be familiar with the log (not ln) and 10^x functions of your calculator. Let's dive in.

Brønsted–Lowry Acids and Bases

An acid in the Brønsted–Lowry theory is an H^+ donor, and a base is an H^+ acceptor. In the Brønsted–Lowry acid-base theory, there is a competition for an H^+. Consider the acid-base reaction between acetic acid, a weak acid, and ammonia, a weak base:

$$CH_3COOH(aq) + NH_3(aq) \leftrightarrows CH_3COO^-(aq) + NH_4^+(aq)$$

Acetic acid donates a proton to ammonia in the forward (left to right) reaction of the equilibrium to form the acetate and ammonium ions. However, in the reverse (right to left) reaction, the ammonium ion donates a proton to the acetate ion to form ammonia and acetic acid. The ammonium ion is acting as an acid and the acetate ion a base. Under the Brønsted–Lowry system, acetic acid (CH_3COOH) and the acetate ion (CH_3COO^-) are a conjugate acid-base pair. The members of a **conjugate acid-base pair** differ by only a single H^+. Ammonia (NH_3) and the ammonium ion (NH_4^+) are also a conjugate acid-base pair. In this reaction, there is a competition for the H^+ between acetate ion and ammonia. To predict on which side the equilibrium will lie, this general rule applies: *the equilibrium will favor (shift toward) the side in which the weakest acid and base are present.*

Strength of Acids and Bases

In Chapter 4, we introduced the concept of acids and bases. Acids and bases may be strong or weak. **Strong acids** completely dissociate in water and **weak acids** only partially dissociate. For example, consider two acids HCl (strong) and CH_3COOH (weak). If we add each to water to form aqueous solutions, the following reactions take place:

$$HCl(aq) + H_2O(l) \rightarrow H_3O^+(aq) + Cl^-(aq)$$
$$CH_3COOH(aq) + H_2O(l) \leftrightarrows H_3O^+(aq) + CH_3COO^-(aq)$$

The first reaction goes to completion—there is no HCl left in solution. The second reaction is an equilibrium reaction—there are appreciable amounts of both reactants and products left in solution.

H^+ and H_3O^+ represent the same chemical species.

There are generally only two strong bases to consider—the hydroxide ion and those species that produce hydroxide ion in aqueous solution. All other bases are weak. **Weak bases** also establish an equilibrium system, as in aqueous solutions of ammonia:

$$NH_3(aq) + H_2O(l) \leftrightarrows OH^-(aq) + NH_4^+(aq)$$

Strong acids include:	Strong bases include:
chloric acid, $HClO_3$	alkali metal (Group 1A) hydroxides
hydrobromic acid, HBr	LiOH, NaOH, KOH, RbOH, and CsOH
hydrochloric acid, HCl	calcium, strontium, and barium hydroxides
hydroiodic acid, HI	$Ca(OH)_2$, $Sr(OH)_2$, and $Ba(OH)_2$
nitric acid, HNO_3	
perchloric acid, $HClO_4$	
sulfuric acid, H_2SO_4	

Assume that any acid or base not on the strong acid and strong base lists is weak.

K_w—the Water Dissociation Constant

Water is **amphoteric**. It will act as either an acid or a base, depending on whether the other species is a base or acid. In pure water, we find the same amphoteric nature. In pure water, a very small amount of proton transfer is taking place:

$$H_2O(l) + H_2O(l) \leftrightarrows H_3O^+(aq) + OH^-(aq)$$

This is commonly written as:

$$H_2O(l) \leftrightarrows H^+(aq) + OH^-(aq)$$

There is an equilibrium constant, called the **water dissociation constant, K_w**, that describes this equilibrium:

$$K_w = [H^+] [OH^-] = 1.00 \times 10^{-14} \text{ at 25 °C}$$

The numerical value for the K_w of 1.00×10^{-14} is for the product of the $[H^+]$ and $[OH^-]$ in pure water and for all aqueous solutions.

In an aqueous solution, there are two sources of H^+, the acid and water. However, the amount of H^+ that the water dissociation contributes is very small and is normally easily ignored.

BTW

The concentration of water (solvent) is a constant and is incorporated into the K_w.

pH

Because the concentration of the hydronium ion, H_3O^+ (or H^+ as a shorthand notation), can vary tremendously in solutions of acids and bases, a scale to represent the acidity of a solution was developed. This is the pH scale, which relates to the $[H_3O^+]$:

pH = $-\log [H_3O^+]$ or $-\log [H^+]$ using the shorthand notation.

Remember that in pure water $K_w = [H^+] [OH^-] = 1.00 \times 10^{-14}$. Since both the hydronium ion and hydroxide ions form in equal amounts, x, the K_w expression is:

$[x]^2 = 1.00 \times 10^{-14}$

> **BTW**
>
> *The value of K_w only applies at or near room temperature. This value, like all equilibrium constants, will vary with temperature.*

Solving for $[x]$ gives us $x = [H^+] = 1.00 \times 10^{-7}$. If you then calculate the pH of pure water:

pH = $-\log [H^+] = -\log [1.00 \times 10^{-7}] = - (-7.00) = 7.00$

The pH of pure water is 7.00 and on the pH scale, this is **neutral**. A solution that has a $[H^+]$ greater than pure water will have a pH *less* than 7.00 and is **acidic**. A solution that has a $[H^+]$ less than pure water will have a pH *greater* than 7.00 and is **basic**.

It is also possible to calculate the pOH of a solution. It is defined as pOH = $-\log [OH^-]$. The pH and the pOH are related:

pH + pOH = pK_w = 14.00 at 25°C

EXAMPLE

▶ Calculate the pH and pOH of a 0.025 M nitric acid solution.

▶ Write the dissociation reaction for nitric acid, a strong acid:

$$HNO_3(aq) \rightarrow H^+(aq) + NO_3^-(aq)$$

▶ Since nitric acid is a strong acid, the contribution to the [H⁺] from the nitric acid will be 0.025 M. The contribution of water to the [H⁺] will be insignificant. Thus, [H⁺] = 0.025 M.

$$pH = -\log [H^+] = -\log [0.025] = -(-1.60) = 1.60$$

$$pOH = 14.00 - pH = 14.00 - 1.60 = 12.40$$

In the previous example, we went from concentration to pH. Now let's reverse that process.

▶ A solution had a pH = 8.75. Calculate the [H⁺] and [OH⁻] of the solution.

▶ We may determine the [H⁺] of the solution from the pH using the inverse log function on our calculator or by using the relationship:

$$[H^+] = 10^{-pH}$$

$$[H^+] = 10^{-8.75} = 1.8 \times 10^{-9}$$

▶ To calculate the [OH⁻] we will use the K_w relationship:
$$K_w = [H_3O^+] [OH^-]$$

$$[OH^-] = \frac{K_w}{[H_3O^+]} = \frac{1.00 \times 10^{-14}}{1.8 \times 10^{-9}} = 5.6 \times 10^{-6}$$

Acid-Base Equilibrium (K_a and K_b)

In Chapter 14, we studied equilibrium constants, the way we represent systems that establish an equilibrium. There are specific equilibrium constants associated with acids and bases.

K_a—Acid Dissociation Constant

Strong acids completely dissociate (ionize) in water. Weak acids partially dissociate and establish an equilibrium system. There is a large range of weak acids based on their ability to donate protons. Consider the general weak acid HA and its reaction when placed in water:

BTW
The HA is not necessarily neutral, and the A does not necessarily have a negative charge.

$$HA(aq) \leftrightarrows H^+(aq) + A^-(aq)$$

We can write an equilibrium constant expression for this system:

$$K_c = \frac{[H^+][A^-]}{[HA]}$$

Since this is the equilibrium constant associated with a weak acid dissociation, this K_c is the **weak acid dissociation constant, K_a**. The K_a expression is:

$$K_a = \frac{[H^+][A^-]}{[HA]}$$

EASY MISTAKE
The [H_2O] is assumed to be a constant and is incorporated into the K_a value. It does not appear in the equilibrium constant expression.

The greater the amount of dissociation, the larger the value of the K_a. The following table shows the K_a values for some common acids:

Acid	K_a Value
HSO_4^-	1.2×10^{-2}
$HClO_2$	1.2×10^{-2}
$HC_2H_2ClO_2$	1.35×10^{-3}
HF	7.2×10^{-4}
HNO_2	4.0×10^{-4}
$HC_2H_3O_2$	1.8×10^{-5}
$[Al(H_2O)_6]^{3+}$	1.4×10^{-5}
$HOCl$	3.5×10^{-8}
HCN	6.2×10^{-10}
NH_4^+	5.6×10^{-10}
HOC_6H_5	1.6×10^{-10}

BTW

For every H^+ formed, there is an A^- formed.

If we know the initial molarity and K_a of the weak acid, we can easily calculate the $[H^+]$ or $[A^-]$. If we know the initial molarity and $[H^+]$, it is possible to calculate the K_a.

For **polyprotic acids**, acids that can donate more than one proton, the K_a for the first dissociation is normally much larger than the K_a for the second dissociation. If there is a third K_a, it is normally much smaller still. For most practical purposes, you can simply use the first K_a.

BTW

The [HA] is the equilibrium molar concentration of the undissociated weak acid, not its initial concentration. The exact expression would then be [HA] = M$_{initially}$ − [H$^+$], where M$_{initially}$ is the initial concentration of the weak acid. This is true because for every H$^+$ that is formed an HA must have dissociated. However, many times if the K$_a$ is small, you can approximate the equilibrium concentration of the weak acid by its initial concentration, [HA] < M$_{initial}$.

K_b—the Base Dissociation Constant

Weak bases (B), when placed into water, also establish an equilibrium system much like weak acids:

$$B(aq) + H_2O(l) \leftrightarrows HB^+(aq) + OH^-(aq)$$

The equilibrium constant expression is the weak **base dissociation constant**, K_b, and has the form:

$$K_b = \frac{[HB^+][OH^-]}{[HB]}$$

The same reasoning that was used in dealing with weak acids is also true here: $[HB^+] = [OH^-] = x$; $[HB] < M_{initial}$; the numerator can be represented as $[x]^2$; and knowing the initial molarity and K_b of the weak base, the $[OH^-]$ can be calculated. If we know the initial molarity and $[OH^-]$, it is possible to calculate the K_b.

The K_a and K_b of conjugate acid-base pairs are related through the K_w expression:

$$K_a K_b = K_w$$

Acid-Base Properties of Salts and Oxides

Some salts have acid-base properties. Ammonium chloride, NH_4Cl, when dissolved in water will dissociate and the ammonium ion will act as a weak acid, donating a proton, which makes the solution acidic.

Certain oxides can have acid or basic properties. Many oxides of metals that have a +1 or +2 charge are basic oxides because they will react with water to form a basic solution:

$$Na_2O(s) + H_2O(l) \rightarrow 2\ NaOH(aq)$$

Many nonmetal oxides are acidic oxides because they react with water to form an acidic solution:

$$Cl_2O_5(s) + H_2O(l) \rightarrow 2\ HClO_3(aq)$$

Another salt-like group of compounds that have acid-base properties is the hydrides of the alkali metals and calcium, strontium, and barium. These hydrides will react with water to form the hydroxide ion and hydrogen gas:

$$KH(s) + H_2O(l) \rightarrow KOH(aq) + H_2(g)$$

Lewis Acids and Bases

Another acid-base theory is the Lewis acid-base theory. According to this theory, a Lewis acid will accept a pair of electrons and a Lewis base will donate a pair of electrons. To make it easier to see which species is donating electrons, it is helpful to use Lewis structures for the reactants and if possible, for the products.

The following is an example of a Lewis acid-base reaction.

$$H^+(aq) + :NH_3(aq) \rightarrow H-NH_3^+(aq)$$

The hydrogen ion accepts the lone pair of electrons (shown on the nitrogen) from the ammonia to form the ammonium ion. The hydrogen ion, because it accepts a pair of electrons, is the Lewis acid. The ammonia, because it donates a pair of electrons, is the Lewis base. This reaction is also a Brønsted–Lowry acid-base reaction. This illustrates that a substance may be an acid or a base by more than one definition. All Brønsted–Lowry acids are Lewis acids, and all Brønsted–Lowry bases are Lewis bases. However, the reverse is not necessarily true.

Working Weak Acid and Base Problems

Problems involving weak acids and bases are really worked the same way as any chemical equilibrium problem (Chapter 14). Let's begin with a series of acid-base equilibria problems.

EXAMPLE

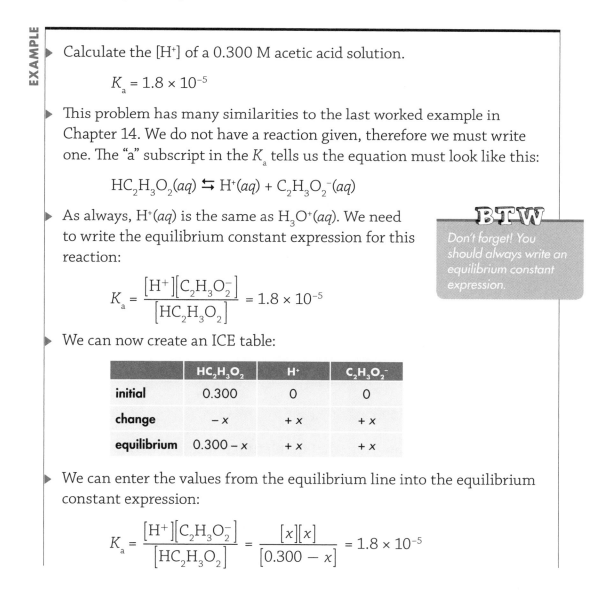

▶ Calculate the [H⁺] of a 0.300 M acetic acid solution.

$$K_a = 1.8 \times 10^{-5}$$

▶ This problem has many similarities to the last worked example in Chapter 14. We do not have a reaction given, therefore we must write one. The "a" subscript in the K_a tells us the equation must look like this:

$$HC_2H_3O_2(aq) \rightleftharpoons H^+(aq) + C_2H_3O_2^-(aq)$$

▶ As always, H⁺(aq) is the same as H₃O⁺(aq). We need to write the equilibrium constant expression for this reaction:

$$K_a = \frac{[H^+][C_2H_3O_2^-]}{[HC_2H_3O_2]} = 1.8 \times 10^{-5}$$

BTW

Don't forget! You should always write an equilibrium constant expression.

▶ We can now create an ICE table:

	HC₂H₃O₂	H⁺	C₂H₃O₂⁻
initial	0.300	0	0
change	− x	+ x	+ x
equilibrium	0.300 − x	+ x	+ x

▶ We can enter the values from the equilibrium line into the equilibrium constant expression:

$$K_a = \frac{[H^+][C_2H_3O_2^-]}{[HC_2H_3O_2]} = \frac{[x][x]}{[0.300 - x]} = 1.8 \times 10^{-5}$$

▶ You could simplify this expression by assuming $0.300 - x = 0.300\,M$ or do a quadratic. The hydrogen ion concentration is $x = [H^+] = 2.3 \times 10^{-3}\,M$. Thus, the solution is acidic.

Now let us try an example using a weak base.

EXAMPLE

▶ A 0.500 M solution of ammonia has a pH of 11.48. What is the K_b of ammonia?

▶ We do not have a reaction given; therefore, we must write one. The "b" subscript in the K_b tells us the equation must look like this:

$$NH_3(aq) + H_2O(l) \rightleftharpoons NH_4^+(aq) + OH^-(aq)$$

▶ We need to write the equilibrium constant expression for this reaction:

$$K_b = \frac{[OH^-][NH_4^+]}{[NH_3]}$$

▶ We can now create an ICE table:

	NH₃	OH⁻	NH₄⁺
initial	0.500	0	0
change	− x	+ x	+ x
equilibrium	0.500 − x	+ x	+ x

▶ We can enter the values from the equilibrium line into the equilibrium constant expression:

$$K_b = \frac{[OH^-][NH_4^+]}{[NH_3]} = \frac{[x][x]}{[0.500 - x]}$$

▶ There is one additional piece of information in the problem, and that is the pH. From the pH we can determine the hydrogen ion concentration:

$$pH = 11.48$$

$$[H^+] = 10^{-11.48}$$

$$[H^+] = 3.3 \times 10^{-12} \, M$$

▶ From the hydrogen ion concentration and K_w, we can determine the hydroxide ion concentration in the solution.

$$K_w = [H^+][OH^-] = 1.00 \times 10^{-14}$$

$$[OH^-] = \frac{K_w}{[H^+]} = \frac{1.00 \times 10^{-14}}{3.3 \times 10^{-12}} = 3.0 \times 10^{-3} \, M$$

▶ From our table, we know that the hydroxide ion concentration is x. Therefore, we can substitute 3.0×10^{-3} into the equilibrium expression for x and enter the values into a calculator:

$$K_b = \frac{[OH^-][NH_4^+]}{[NH_3]} = \frac{[x][x]}{[0.500 - x]} = \frac{[3.0 \times 10^{-3}][3.0 \times 10^{-3}]}{[0.500 - 3.0 \times 10^{-3}]} = 1.8 \times 10^{-5}$$

EXTRA HELP

In our final example, we will do a problem with a twist.

▶ Determine the pH of a 0.200 M strontium acetate solution.

▶ Strontium acetate is neither a weak acid nor a weak base—it is a soluble salt. As a soluble salt, it is a strong electrolyte, and it will dissociate as follows:

$$Sr(C_2H_3O_2)_2(aq) \rightarrow Sr^{2+}(aq) + 2\,C_2H_3O_2^-(aq)$$

▶ The resulting solution has 0.200 M Sr^{2+} (we do not need this value) and $2(0.200\ M) = 0.400\ M\ C_2H_3O_2^-$.

▶ We can ignore ions such as Sr^{2+}, which come from strong acids or strong bases in this type of problem. Ions, such as $C_2H_3O_2^-$, from a weak acid or a base, weak acid in this case, will undergo hydrolysis, a reaction with water. The acetate ion is the conjugate base of acetic acid ($K_a = 1.74 \times 10^{-5}$). Since acetate is a weak base, this will be a K_b problem, and OH^- will form. The equilibrium is:

$$C_2H_3O_2^-(aq) + H_2O(l) \rightleftharpoons OH^-(aq) + HC_2H_3O_2(aq)$$

▶ The equilibrium constant expression for this reaction is:

$$K_b = \frac{[OH^-][HC_2H_3O_2]}{[C_2H_3O_2^-]}$$

▶ We can now create an ICE table:

	$C_2H_3O_2^-$	OH^-	$HC_2H_3O_2$
initial	0.400	0	0
change	$-x$	$+x$	$+x$
equilibrium	$0.400 - x$	$+x$	$+x$

▶ We can enter the values from the equilibrium line into the equilibrium constant expression:

$$K_b = \frac{[OH^-][HC_2H_3O_2]}{[C_2H_3O_2^-]} = \frac{[x][x]}{[0.400 - x]}$$

▶ We need to determine the value of K_b to finish the problem. We know K_a and we know that $K_a K_b = K_w = 1.0 \times 10^{-14}$. Therefore, we can find K_b from:

$$K_b = \frac{K_w}{K_a} = \frac{1.00 \times 10^{-14}}{1.74 \times 10^{-5}} = 5.7471 \times 10^{-10} \text{ (unrounded)}$$

▶ We can add this information to the equilibrium constant expression:

$$K_b = \frac{[OH^-][HC_2H_3O_2]}{[C_2H_3O_2^-]} = \frac{[x][x]}{[0.400 - x]} = 5.7471 \times 10^{-10}$$

▶ From this equation, we can determine that $x = 1.51619 \times 10^{-5}$ M (unrounded). This is the hydroxide ion concentration.

▶ There are two ways to determine the pH from the hydroxide ion concentration. One method uses K_w, while the other method uses the pOH of the solution. The K_w approach is:

$$K_w = [H^+][OH^-] = 1.0 \times 10^{-14}$$

$$[H^+] = \frac{K_w}{[OH^-]} = \frac{1.00 \times 10^{-14}}{1.51619 \times 10^{-5}} = 6.5955 \times 10^{-10} \text{ (unrounded)}$$

$$pH = -\log[H^+] = -\log(6.5955 \times 10^{-10}) = 9.18075 = 9.18$$

▶ The pOH method begins by determining the pOH of the solution:

$$pOH = -\log[OH^-] = -\log(1.51619 \times 10^{-5}) = 4.8192$$
(unrounded)

▶ Next, we use the relationship $pK_w = pH + pOH = 14.00$.

$$pH = pK_w - pOH = 14.00 - 4.8192 = 9.18$$

▶ Both methods give us a basic solution, which we should expect from a base like the acetate ion.

BTW

*The pH of any acid solution **must** be below 7 and the [H⁺] must be greater than 10⁻⁷ M. The pH of any base solution **must** be above 7 and the [OH⁻] must be greater than 10⁻⁷ M ([H⁺] < 10⁻⁷ M). If your results disagree with this, you have made an error.*

EXERCISES

EXERCISE 15-1

Answer the following questions.

1. List the strong acids.

2. List the strong bases.

3. What is the relationship between the K_a and the K_b for any conjugate acid-base pair?

4. What mathematical relationship defines the pH of a solution?

5. What is the relationship between the pH and the pOH for any aqueous solution?

6. Determine the $[OH^-]$, pOH, and pH of a 0.100 M solution of ammonia, NH_3. $K_b = 1.8 \times 10^{-5}$.

EXERCISE 15-2

Give the formula for the conjugate base of each of the following.

1. HNO_3

2. $HC_2H_3O_2$

3. NH_3

4. H_2SO_4

5. HCO_3^-

EXERCISE 15-3

Give the formula for the conjugate acid of each of the following.

1. Cl^-

2. PO_4^{3-}

3. NH_3

4. NO_2^-

5. O^{2-}

EXERCISE 15-4

Determine the pH in each of the following.

1. $[H^+] = 10^{-3}$ M

2. $[H^+] = 1.5 \times 10^{-5}$ M

3. $[H^+] = 1.0$ M

4. $[H^+] = 1.0 \times 10^{-15} M$

5. $pOH = 4.2$

EXERCISE 15-5

Determine the hydrogen ion concentration in each of the following.

1. $pH = 4$

2. $pH = 3.75$

3. $pOH = 7.0$

4. $[OH^-] = 3.2 \times 10^{-3} M$

5. $pH = -0.50$

EXERCISE 15-6

Determine the pH of each of the following.

1. $1.0 \times 10^{-2}\ M$ HCl

2. $4.5 = 10^{-5}\ M$ HNO_3

3. $1.5 \times 10^{-3}\ M$ NaOH

EXERCISE 15-7

Determine the [H⁺] of each of the following.

1. $1.0 \times 10^{-1}\ M$ $HC_2H_3O_2$, $K_a = 1.8 \times 10^{-5}$

2. $4.5 \times 10^{-2}\ M$ $HClO_2$, $K_a = 1.1 \times 10^{-2}$

3. $1.5\ M$ HCO_3^-, $K_a = 4.8 \times 10^{-11}$

EXERCISE 15-8

Determine the pH of each of the solutions in exercise 15-7.

Flashcard
App

Buffers and Additional Equilibria

MUST KNOW

- The common-ion effect is an application of Le Châtelier's principle to equilibrium systems.

- A buffer is a solution that resists a change in pH if we add an acid or base.

- We use titrations to determine the concentration of an acid or base solution.

e're going to continue learning about acid-base equilibrium systems and buffers and titrations. If you are a little unsure about equilibria and especially weak acid-base equilibria, review Chapters 14 and 15. We will also learn to apply the basic concepts of equilibria to solubility and complex ions. Two things to remember: (1) The basic concepts of equilibria apply to *all* the various types of equilibria, and (2) the more problems you work, the better your understanding.

The Common Ion Effect

If a slightly soluble salt solution is at equilibrium and we add a solution containing one of the ions involved in the equilibrium, the solubility of the slightly soluble salt decreases. For example, consider the $PbSO_4$ equilibrium:

$$PbSO_4(s) \leftrightarrows Pb^{2+}(aq) + SO_4^{2-}(aq)$$

Suppose we add a solution of Na_2SO_4 to this equilibrium system. The additional sulfate ion will disrupt the equilibrium by Le Châtelier's principle and shift it to the left. This decreases the solubility. The same would be true if you tried to dissolve $PbSO_4$ in a solution of Na_2SO_4 instead of pure water— the solubility would be less. This application of Le Châtelier's principle to equilibrium systems of a slightly soluble salt is the **common-ion effect**. In the next section we are going to consider an application of the common-ion effect—buffers.

BTW

A slightly soluble salt is any salt predicted to be "insoluble" by the solubility rules. Nothing is completely insoluble.

Buffers and pH

Buffers are solutions that resist a change in pH when we add an acid or base. A buffer contains both a weak acid (HA) and its conjugate base (A^-). The acid part will neutralize any base added and the base part of the buffer will neutralize any acid added to the solution. We may calculate the hydronium ion concentration of a buffer by rearranging the K_a expression to yield the **Henderson–Hasselbalch equation**, which we can use to calculate the pH of a buffer:

$$pH = pK_a + \log\frac{[A^-]}{[HA]}$$

or using the K_b expression:

$$pOH = pK_b + \log\frac{[HA]}{[A^-]}$$

Buffer solutions are also examples of the common-ion effect applied to a K_a or K_b equilibrium.

These equations allow us to calculate the pH of the buffer solution knowing the K_a of the weak acid or K_b of the weak base and the concentrations of the weak acid and its conjugate base. If we know the desired pH along with the K_a of the weak acid, then the ratio of base to acid can be calculated. The more concentrated the acid and base components are, the more acid or base can be neutralized and the less the change in buffer pH. This is a measure of the **buffer capacity**, the ability to resist a change in pH.

 IRL Living organisms are very sensitive to a change in pH. A change of as little as 0.2 pH can cause medical distress or death. There are several acid-base buffer systems in the blood that work to keep the blood pH within safe limits.

Let's begin by determining the pH of a buffer solution.

What is the pH of a solution containing 2.00 mol of ammonia and 3.00 mol of ammonium chloride in a volume of 1.00 L? $K_b = 1.81 \times 10^{-5}$?

$$NH_3(aq) + H_2O(l) \rightleftarrows NH_4^+(aq) + OH^-(aq)$$

There are two ways to solve this problem. Assume $x = [OH^-] = [NH_4^+]$ that comes from the reaction of the ammonia. Ammonia is the conjugate base of the ammonium ion. The 2.00 mol of ammonia and the 3.00 mol of ammonium chloride in 1.00 L gives an initial concentration of 2.00 M and 3.00 M, respectively.

$$K_b = \frac{[OH^-][NH_4^+]}{[NH_3]} = \frac{[3.00 + x][x]}{[2.00 - x]} = 1.81 \times 10^{-5}$$

Assume x is small to allow us to bypass the quadratic:

$$1.81 \times 10^{-5} = \frac{[3.00][x]}{[2.00]}$$

$$x = 1.21 \times 10^{-5} \quad \text{and} \quad pOH = 4.918$$

$$\downarrow$$

$$pH = pK_w - pOH = 14.000 - 4.918 = 9.082$$

Here is an alternate solution:

$$pOH = pK_b + \log\frac{[NH_4^+]}{[NH_3]} \quad (pK_b = -\log K_b)$$

$$= 4.742 + \log\frac{[3.00]}{[2.00]}$$

$$\downarrow$$

$$pOH = 4.918 \quad pH = 9.082$$

Notice that using the Henderson–Hasselbalch equation not only got the same answer but did so with less work.

Titrations and Indicators

An acid-base **titration** is a laboratory procedure that we use to determine the concentration of an unknown solution. We add a base solution of known concentration to an acid solution of unknown concentration (or vice versa) until an acid-base **indicator** visually signals that the **endpoint** of the titration has been reached. The **equivalence point** is the point at which we have added a stoichiometric amount of the base to the acid.

If the acid being titrated is a weak acid, then there are equilibria that will be established and accounted for in the calculations. (See the pH and titration problems section at the end of the chapter.) Typically, a plot of pH of the weak acid solution being titrated versus the volume of the strong base added (the **titrant**) starts at a low pH and gradually rises until close to the equivalence point in which the curve rises dramatically. After the equivalence point region, the curve returns to a gradual increase. We can see this in the following figure.

Titration of a Weak Acid with a Strong Base

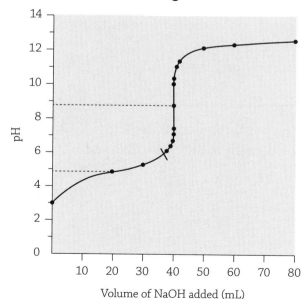

Solubility Equilibria (K_{sp})

Many salts are soluble in water, but others are only slightly soluble. These salts, when placed in water, quickly reach their solubility limit and the ions establish an equilibrium system:

$$PbSO_4(s) \leftrightarrows Pb^{2+}(aq) + SO_4^{2-}(aq)$$

The equilibrium constant expression associated with systems of slightly soluble salts is the **solubility product constant, K_{sp}**. It is the product of the ionic concentrations, each one raised to the power of the coefficient in the balanced chemical equation. It contains no denominator since the concentration of a solid is, by convention, 1, and for this reason it does not appear in the equilibrium constant expression. The K_{sp} expression for the $PbSO_4$ system is:

$$K_{sp} = [Pb^{2+}] [SO_4^{2-}]$$

For this salt, the value of the K_{sp} is 1.6×10^{-8} at 25°C. If we know the value of the solubility product constant, then we can determine the concentration of the ions. In addition, if we know one of the ion concentrations, then we can find the K_{sp}.

Let's try a K_{sp} example problem.

EXAMPLE

The K_{sp} of magnesium fluoride in water is 8×10^{-8}. How many grams of magnesium fluoride will dissolve in 0.250 L of water?

$$MgF_2(s) \leftrightarrows Mg^{2+}(aq) + 2 F^-(aq)$$

$$K_{sp} = [Mg^{2+}] [F^-]^2 = 8 \times 10^{-8}$$

▶ For every mole of MgF_2 that dissolves, one mole of Mg^{2+} and two moles of F^- form:

$$K_{sp} = (x)(2x)^2 = 4x^3 = 8 \times 10^{-8}$$

$$x = 3 \times 10^{-3} = [Mg^{2+}]$$

$$\downarrow$$

$$\left(\frac{3 \times 10^{-3} \text{ mol } Mg^{2+}}{L}\right)(0.250 \text{ L})\left(\frac{1 \text{ mol } MgF_2}{1 \text{ mol } Mg^{2+}}\right)\left(\frac{62.3 \text{ g } MgF_2}{1 \text{ mol } MgF_2}\right) = 0.05 \text{ g}$$

When solving common-ion-effect problems, calculations like the preceding ones involving finding concentrations and K_{sp} values can still be done; however, the concentration of the additional common ion will have to be inserted into the solubility product constant expression. Sometimes, if the K_{sp} is very small and the common ion concentration is large, we can simply approximate the concentration of the common ion by the concentration of the ion added.

EXAMPLE

▶ Calculate the silver ion concentration in each of the following solutions:

    a. $Ag_2CrO_4(s)$ + water

    b. $Ag_2CrO_4(s)$ + 1.00 M Na_2CrO_4

▶ For Ag_2CrO_4 $K_{sp} = 1.9 \times 10^{-12}$

    a. $Ag_2CrO_4(s) \leftrightarrows 2 Ag^+ + CrO_4^{2-}$

                         $2x$ x

$$K_{sp} = 1.9 \times 10^{-12} = (2x)^2(x) = 4x^3$$

$$x = 7.8 \times 10^{-5}$$

$$\downarrow$$

$$[Ag^+] = 2x = 1.6 \times 10^{-4} \text{ } M$$

b. $1.00\ M\ Na_2CrO_4 \rightarrow 1.00\ M\ CrO_4^{2-}$ (common ion)

$$Ag_2CrO_4(s) \leftrightarrows 2\ Ag^+ + CrO_4^{2-}$$
$$\ 2x \quad\ 1.00 + x$$

$$K_{sp} = 1.9 \times 10^{-12} = (2x)^2(1.00 + \cancel{x}) = 1.00\ (4x^2) = 4x^2$$

$$\text{neglecting} + x$$

$$x = 6.9 \times 10^{-7}$$

$$\downarrow$$

$$[Ag^+] = 2x = 1.4 \times 10^{-6}\ M$$

Knowing the value of the solubility product constant can also allow us to predict whether a precipitate will form if we mix two solutions, each containing an ion component of a slightly soluble salt. We calculate the **reaction quotient** (many times called the **ion product**), which has the same form as the solubility product constant. We take into consideration the mixing of the volumes of the two solutions, and then compare this reaction quotient to the K_{sp}. If it is greater than the K_{sp} then precipitation will occur until the ion concentrations reduce to the solubility level.

EXAMPLE

▶ If 10.0 mL of a 0.100 M $BaCl_2$ solution are added to 40.0 mL of a 0.0250 M Na_2SO_4 solution, will $BaSO_4$ precipitate? K_{sp} for $BaSO_4$ = 1.1×10^{-10}.

▶ To answer this question, we will need the concentration of the barium ion and the sulfate ion *before* precipitation. These may be determined from the dilution relationship:

$$M_{dil} = M_{con}V_{con} / V_{dil}$$

For Ba^{2+}: M_{dil} = (0.100 M) (10.0 mL) / (10.0 + 40.0) mL
$$= 0.0200\ M$$

For SO_4^{2-}: M_{dil} = (0.0250 M) (40.0 mL) / (50.0 mL) = 0.0200 M

▶ Inserting these values into the reaction quotient relationship gives:

$$Q = [Ba^{2+}] [SO_4^{2-}] = (0.0200)(0.0200) = 0.000400 = 4.00 \times 10^{-4}$$

↓

$$4.00 \times 10^{-4} > 1.1 \times 10^{-10}$$

▶ Since Q is greater than K_{sp}, precipitation will occur.

K_f—Complex Ion Equilibria

We can treat other types of equilibria in much the same way as the ones previously discussed. For example, there is an equilibrium constant associated with the formation of complex ions. This equilibrium constant is the **formation constant, K_f**.

$[Zn(H_2O)_4]^{2+}$ reacts with ammonia to form the $[Zn(NH_3)_4]^{2+}$ complex ion according to the following equation:

$$[Zn(H_2O)_4]^{2+}(aq) + 4 NH_3(aq) \leftrightarrows [Zn(NH_3)_4]^{2+}(aq) + 4 H_2O(l)$$

The K_f of $[Zn(NH_3)_4]^{2+}(aq)$ is 7.8×10^8, indicating that the equilibrium lies to the right, because the value is very large.

EXTRA HELP

pH and Titrations Problems

Let's examine a typical titration problem. A typical chemistry problem may use only one of the steps we will see in this problem, or you may need to work all these steps. The goal of each step is to determine the pH of the solution. You should be careful not to lose sight of this goal.

In many cases, you may know the initial concentration of the weak acid but may be interested in the pH changes during the titration. To do this, you can divide the titration curve into four distinctive areas in which the pH is calculated.

1. This first section is the initial pH. This is the point in the titration preceding the addition of any reactant. You must ignore the reactant to be added in this region. If a strong acid or a strong base is present, there are no significant equilibria. The strong acid gives the hydrogen ion concentration directly, while a strong base gives the hydroxide ion concentration directly. If a weak acid is present, this is a generic K_a problem, and you can find the hydrogen ion concentration from this. If a weak base is present, this is a generic K_b problem, and you can find the hydroxide ion concentration from this. The hydrogen ion concentration directly gives you the pH, while the hydroxide ion concentration will give you the pH indirectly.

   In all other portions of the titration curve, you must consider both the substance already in the container and the amount added. Calculations in these regions will begin as a limiting reagent problem. The results of the limiting reagent calculation will tell you how to finish the problem.

   To do a limiting reagent problem, you need to know how many moles of reactant are in the original container and how many moles of the other reactant have been added. Usually you will find the moles by multiplying the concentration by the volume of the solution. In a few cases, you may find the number of moles from the mass and the molar mass.

2. In the second region of the titration curve, the substance added is the limiting reagent. As a limiting reagent, it is no longer present to affect the pH. The other reactant is in excess, and this excess will influence the pH. The moles of excess reactant divided by the total volume of the solution (the initial volume plus the volume of the solution added) gives the concentration. The solution also contains the products of the reaction. If the excess reactant is either a weak acid or a weak base, the concentration of the products will be important. A buffer solution will be present any time you have an excess of a weak acid or weak base along with its conjugate.

3. The next point in the titration curve is the equivalence point. At this point, both the material added and the material originally present are limiting. At this point, neither of the reactants will be present and therefore will not affect the pH. If the titration involves a strong acid and a strong base, the pH at the equivalence point is 7. If the titration involves a weak base, only the conjugate acid is present to affect the pH. This will require a K_a calculation, and the pH will be acidic. If the titration involves a weak acid, only the conjugate base is present to affect the pH. This will require a K_b calculation, and the pH will be basic. The calculation of the conjugate acid or base will be the moles produced divided by the total volume of the solution.

4. The final region of the titration curve is after the equivalence point. In this region, the material originally present in the container is limiting. The excess reagent, the material added, will affect the pH. If this excess reactant is a weak acid or a weak base, this will be a buffer solution.

   Let's use the following example to see how to deal with each of these regions. This will be a titration of the weak acid with a strong base.

EXAMPLE

A 100.0 mL sample of 0.150 M nitrous acid (pK_a = 3.35) was titrated with 0.300 M sodium hydroxide, NaOH. Determine the pH of the solution after the following quantities of base have been added to the acid solution:

  a. 0.00 mL

  b. 25.00 mL

  c. 50.00 mL

  d. 55.00 mL

a. 0.00 mL   Since no NaOH has been added, **this is the initial pH section of the titration curve**.

▶ The only substance present is nitrous acid, HNO_2, a weak acid. Since this is a weak acid, this must be a K_a problem. As a K_a problem, we can set up a simple equilibrium problem. The K_a expression is:

BTW

For simple equilibria, such as this one, it is possible to use this abbreviated ICE table (only the last line of the ICE table).

$$HNO_2(aq) \leftrightarrows H^+(aq) + NO_2^-(aq)$$

$$0.150 - x \qquad x \qquad x$$

$$K_a = \frac{[H^+][NO_2^-]}{[HNO_2]}$$

▶ We can now enter the concentrations of the various substances:

$$K_a = \frac{[H^+][NO_2^-]}{[HNO_2]} = \frac{[x][x]}{[0.150 - x]}$$

▶ The next step is to find the K_a from the pK_a.

$$K_a = 10^{-pK_a} = 10^{-3.35} = 4.4668 \times 10^{-4} \text{ (unrounded)}$$

$$K_a = \frac{[H^+][NO_2^-]}{[HNO_2]} = \frac{[x][x]}{[0.150 - x]} = 4.4668 \times 10^{-4}$$

▶ Solving this relationship for x gives $x = 8.1855 \times 10^{-3}$ M (unrounded). This is the hydrogen ion concentration, so we can use it to determine the pH:

$$pH = -\log [H^+] = -\log (8.1873 \times 10^{-3}) = 2.08686 = 2.09$$

▶ The initial pH of this weak acid is, as expected, acidic.

▶ **b.** 25.00 mL This will be in the **second region of the titration curve**. (We usually cannot predict this by simple inspection.) To do this, and all later steps, we will require the balanced chemical equation for the reaction and the moles of each reactant.

▶ The reaction is:

$$HNO_2(aq) + NaOH(aq) \rightarrow Na^+(aq) + NO_2^-(aq) + H_2O(l)$$

▶ We need to deal with the stoichiometry of this reaction. For this reason, we need to know the moles of each of the reactants. We can find these from the concentration and the volume of each solution.

BTW

It is usually helpful to write the products in ionic form.

$$\text{mole HNO}_2 = \left(\frac{0.150 \text{ mol HNO}_2}{L}\right)\left(\frac{1 \text{ L}}{1000 \text{ mL}}\right)(100.00 \text{ mL})$$

$$= 0.0150 \text{ mol HNO}_2$$

$$\text{moles NaOH} = \left(\frac{0.300 \text{ mol NaOH}}{L}\right)\left(\frac{1 \text{ L}}{1000 \text{ mL}}\right)(25.00 \text{ mL})$$

$$= 0.00750 \text{ mol NaOH}$$

▶ We will need the moles of HNO_2 in all the remaining steps in this problem. The moles of NaOH will be changing as we add more. The coefficients in the reaction are all ones; thus, we can simply compare the moles to find the limiting reactant. The sodium hydroxide, with the smaller number of moles, is limiting. We can add the mole information to the balanced chemical equation. (The water, being neutral, will not be tracked.)

	$HNO_2(aq)$ +	$NaOH(aq) \rightarrow$	$Na^+(aq)$ +	$NO_2^-(aq)$ +	$H_2O(l)$
initial moles	0.0150	0.00750	0	0	—

▶ The reaction will decrease the moles of reactants present and increase the moles of products. This leads to a reacted line in the table. In this line, we subtract the moles of limiting reactant from each reactant and add the moles to each product.

	HNO$_2$(aq) +	NaOH(aq) →	Na$^+$(aq) +	NO$_2^-$(aq) +	H$_2$O(l)
initial moles	0.0150	0.00750	0	0	—
reacted moles	−0.00750	−0.00750	+0.00750	+0.00750	

Adding each column in this table gives us the post–reaction amounts:

	HNO$_2$(aq) +	NaOH(aq) →	Na$^+$(aq) +	NO$_2^-$(aq) +	H$_2$O(l)
initial moles	0.0150	0.00750	0	0	—
reacted moles	−0.00750	−0.00750	+0.00750	+0.00750	
post-reaction	0.0075	0.00000	0.00750	0.00750	

After the reaction, the moles of sodium hydroxide, limiting reagent, are zero; therefore, sodium hydroxide no longer affects the pH. The solution only contains unreacted nitrous acid, sodium ions, nitrite ions, and water. Cations coming from a strong base do not affect the pH, so we do not need to worry about the sodium ions. Water will not affect the pH either. The nitrous acid and its conjugate base, the nitrite ion, NO$_2^-$, will influence the pH. Nitrous acid and the nitrite ion are a conjugate acid-base pair of a weak acid, and the presence of both in the solution makes the solution a buffer. We need to finish the stoichiometry part of the problem by determining the molarity of each of these substances.

EASY MISTAKE

If any of the reactants or products has a coefficient other than one, you should multiply the moles added or subtracted by this coefficient.

$$M \text{ HNO}_2 = \left(\frac{0.0075 \text{ mol HNO}_2}{(100.00 + 25.00) \text{ mL}} \right) \left(\frac{1000 \text{ mL}}{\text{L}} \right) = 0.060 \ M \text{ HNO}_2$$

$$M\ NO_2^- = \left(\frac{0.0075\ \text{mol}\ NO_2^-}{(100.00 + 25.00)\ \text{mL}}\right)\left(\frac{1000\ \text{mL}}{L}\right) = 0.0600\ M\ NO_2^-$$

▶ We can now use these two values for the equilibrium portion of the problem. There are two options for this buffer solution. We can use these concentrations in a K_a calculation, or we can use the Henderson–Hasselbalch equation. Either method will give you the same answer; however, the Henderson–Hasselbalch equation is faster.

▶ The Henderson–Hasselbalch equation is:

$$pH = pK_a + \log\frac{[A^-]}{[HA]}$$

▶ The pK_a appears in the problem (3.35), HA is nitrous acid, and A^- is the nitrite ion. We can now enter the appropriate values into this equation:

$$pH = 3.35 + \log\frac{[0.0600]}{[0.060]} = 3.35$$

> **BTW**
>
> *When the titration is at the halfway point (concentration of the conjugate acid equals the concentration of the conjugate base), the pH will equal the pK_a, and the pOH will equal the pK_b.*

▶ This is the pH for this point in the second region. All other calculations in this region work this way.

▶ **c.** 50.00 mL   This will be in the **third region of the titration curve**. (We usually cannot predict this by simple inspection.)

▶ We must again use the reaction:

$$HNO_2(aq) + NaOH(aq) \rightarrow Na^+(aq) + NO_2^-(aq) + H_2O(l)$$

▶ We need to deal with the stoichiometry of this reaction. For this reason, we need to know the moles of each of the reactants. We already found the initial moles of nitrous acid (0.0150 mol), so we do not need to determine them again. We can find these from the concentration and the volume of each solution.

$$\text{moles NaOH} = \left(\frac{0.300 \text{ mol NaOH}}{\text{L}}\right)\left(\frac{1 \text{ L}}{1000 \text{ mL}}\right)(50.00 \text{ mL})$$
$$= 0.0150 \text{ mol NaOH}$$

▶ We again need to determine the limiting reagent. In this case, both reactants are limiting. We will now create a new reaction table:

	HNO$_2$(aq) +	NaOH(aq) →	Na$^+$(aq) +	NO$_2^-$(aq) +	H$_2$O(l)
initial moles	0.0150	0.0150	0	0	—
reacted moles	−0.0150	−0.0150	+0.0150	+0.0150	
post-reaction	0.0000	0.00000	0.00150	0.0150	

▶ When both reactants are limiting (zero moles remaining), we are at the equivalence point. The only substance remaining in the solution that can influence the pH is the nitrite ion. This ion is the conjugate base of a weak acid. Since a base is present, the pH will be above 7. The presence of this weak base means this is a K_b problem. However, before we can attack the equilibrium portion of the problem, we must finish the stoichiometry part by finding the concentration of the nitrite ion.

$$\text{M NO}_2^- = \left(\frac{0.0150 \text{ mol NO}_2^-}{(100.00 + 50.00) \text{ mL}}\right)\left(\frac{1000 \text{ mL}}{\text{L}}\right) = 0.100 \; M \; \text{NO}_2^-$$

▶ We can now proceed to the equilibrium portion of the problem. To do this we need the K_b. We can find the K_b from the following series of relationships involving the given pK_a:

$$pK_w = pK_a + pK_b = 14.00$$

$\downarrow$

$$pK_b = pK_w - pK_a = 14.00 - 3.35 = 10.65$$

$$pK_b = -\log K_b = 10.65$$

$\downarrow$

$$pK_b = 10^{-pK_b} = 10^{-10.65} = 2.2387 \times 10^{-11} \text{ (unrounded)}$$

▶ The K_b equilibrium reaction is:

$$NO_2^-(aq) + H_2O(l) \leftrightharpoons OH^-(aq) + HNO_2(aq)$$

▶ This leads to the K_b relationship:

$$K_b = \frac{[OH^-][HNO_2]}{[NO_2^-]} = 2.2387 \times 10^{-11}$$

▶ The ICE table for this equilibrium is:

	$NO_2^-(aq)$	$OH^-(aq)$	$HNO_2(aq)$
initial	0.100	0	0
change	$-x$	$+x$	$+x$
equilibrium	$0.100 - x$	x	x

▶ We can now enter the equilibrium line of the table into the K_b expression and solve for x:

$$K_b = \frac{[OH^-][HNO_2]}{[NO_2^-]} = \frac{[x][x]}{[0.100 - x]} = 2.2387 \times 10^{-11}$$

$\downarrow$

$$x = 1.4962 \times 10^{-6} \; M \; OH^- \text{ (unrounded)}$$

▶ Either we can use the hydroxide ion with the K_w to find the hydrogen ion concentration, and then find the pH, or we can find the pOH and use the pK_w to find the pH. We will use the latter method:

$$pOH = -\log [OH^-] = -\log (1.4962 \times 10^{-6}) = 5.8250 \text{ (unrounded)}$$

$$pK_w = pH + pOH = 14.00$$

$\downarrow$

$$pH = pK_w - pOH = 14.00 - 5.8250 = 8.1750 = 8.18$$

▶ **d.** 55.00 mL 　 This must be in the **fourth region of the titration curve**. (We know this is true because the preceding part gave us the equivalence point.) After the equivalence point, the substance added will be excess and the other substance is limiting. In this case, the sodium hydroxide will be in excess and the nitrous acid will be limiting.

▶ We must again use the reaction:

$$HNO_2(aq) + NaOH(aq) \rightarrow Na^+(aq) + NO_2^-(aq) + H_2O(l)$$

▶ We need to deal with the stoichiometry of this reaction. For this reason, we need to know the moles of each of the reactants. We already found the initial moles of nitrous acid (0.0150 mol), so we do not need to determine them again. We can find these from the concentration and the volume of each solution.

$$\text{moles NaOH} = \left(\frac{0.300 \text{ mol NaOH}}{L} \right)\left(\frac{1 \text{ L}}{1000 \text{ mL}} \right)(55.00 \text{ mL})$$

$$= 0.0165 \text{ mol NaOH}$$

▶ We again need to determine the limiting reagent. In this case, both reactants are limiting. We will now create a new reaction table:

	HNO$_2$(aq) +	NaOH(aq) →	Na$^+$(aq) +	NO$_2^-$(aq) +	H$_2$O(l)
initial moles	0.0150	0.0165	0	0	—
reacted moles	−0.0150	−0.0150	+0.0150	+0.0150	
post-reaction	0.0000	0.0015	0.00150	0.0150	

▶ There are two bases present after the reaction. Both could influence the pH. However, since the sodium hydroxide is a strong base, it will be more important than the weaker base, the nitrite ion. We need to determine the sodium hydroxide ion concentration after the reaction.

$$M \text{ NaOH} = \left(\frac{0.0015 \text{ mol NaOH}}{(100.00 + 55.00) \text{ mL}} \right)\left(\frac{1000 \text{ mL}}{\text{L}} \right)$$
$$= 9.677 \times 10^{-3} \ M \text{ NaOH (unrounded)}$$

▶ Sodium hydroxide is a strong base so M NaOH = M OH$^-$. We can use the hydroxide ion concentration to determine the pH:

$$\text{pOH} = - \log [\text{OH}^-] = - \log (9.677 \times 10^{-3}) = 2.01426 \text{ (unrounded)}$$

$$pK_w = \text{pH} + \text{pOH} = 14.00$$

↓

$$\text{pH} = pK_w - \text{pOH} = 14.00 - 2.01426 = 11.98574 = 11.99$$

▶ All other calculations in this region work in a similar manner.

EXERCISES

EXERCISE 16-1

Answer the following questions.

1. Write the Henderson–Hasselbalch equation for pH and for pOH.

2. Complete the following table for use with buffer solutions.

[HA]	[A⁻]
HNO_2	
	NH_3
HCO_3^-	
	HPO_4^{2-}
$CH_3NH_3^+$	

3. Write an equilibrium constant expression (K_f) for the formation of the $Ag(NH_3)_2^+$ ion.

EXERCISE 16-2

Write equilibrium constant expressions (K_{sp}) for the addition of each of the following substances to water.

1. AgBr

2. Ag_2SO_4

3. AlF_3

4. $CaCO_3$

5. $Ca_3(PO_4)_2$

EXERCISE 16-3

Calculate the molar solubility of calcium carbonate, $CaCO_3$, that will dissolve in each of the following. The K_{sp} for calcium carbonate is 8.7×10^{-9}.

1. in pure water

2. in $6.5 \times 10^{-2} \ M \ Ca^{2+}$

3. in $0.35 \ M \ CO_3^{2-}$ solution

EXERCISE 16-4

Determine the pH of each of the following buffer solutions.

1. $0.15 \ M \ HC_2H_3O_2$ and $0.15 \ M \ C_2H_3O_2^-$ with $pK_a = 4.76$

2. $0.25 \ M \ NH_3$ and $0.30 \ M \ NH_4^+$ with $pK_b = 4.76$

3. $1.0 \ M \ HF$ and $0.75 \ M \ F^-$ with $K_a = 6.8 \times 10^{-4}$

4. $1.2 \ M \ C_5H_5N$ and $0.75 \ M \ HC_5H_5N^+$ with $K_b = 1.5 \times 10^{-9}$

5. $0.60 \ M \ HCO_3^-$ and $0.80 \ M \ CO_3^{2-}$ with $pK_{a1} = 6.35$ and $pK_{a2} = 10.32$

EXERCISE 16-5

The pH of a 50.00 mL sample of 0.06000 M strontium hydroxide is measured as it is titrated with a 0.1200 M hydrochloric acid. Determine the pH of the solution after the following total volumes of hydrochloric acid have been added.

1. 0.00 mL

2. 25.00 mL

3. 50.00 mL

4. 75.00 mL

EXERCISE 16-6

The pH of a 50.00 mL sample of 0.1100 M ammonia, NH_3, is measured as it is titrated with a 0.1100 M hydrochloric acid, HCl. Determine the pH of the solution after the following total volumes of hydrochloric acid have been added.

1. 0.00 mL

2. 25.00 mL

3. 50.00 mL

4. 75.00 mL (K_b for NH_3 is 1.76×10^{-5})

17 Entropy and Free Energy

MUST KNOW

 The first law of thermodynamics states that the total energy of the universe is constant.

 The second law of thermodynamics states that, in all spontaneous processes, the entropy increases.

The Entropy is a measure of the dispersion of energy from a localized one to a more disperse one.

Gibbs free energy is the best indicator of a spontaneous reaction.

e're now going to investigate the laws of thermodynamics, especially the concepts of entropy and free energy. It might be helpful to review Chapter 6 on thermochemistry and the writing of thermochemical equations. Just like in all the previous chapters, to do well you must work a bunch of problems. Remember, to do well, you must work a bunch of problems!

The Three Laws of Thermodynamics

The first law of thermodynamics states that the total energy of the universe is constant. This is simply the law of conservation of energy. We can state this relationship as:

$$\Delta E_{universe} = \Delta E_{system} + \Delta E_{surroundings} = 0$$

The second law of thermodynamics involves a term called *entropy*. **Entropy** is a measure of the degree that energy disperses from a localized state to one that is more widely spread out. We may also think of entropy (S) as a measure of the disorder of a system. The **second law of thermodynamics** states that all processes that occur spontaneously move in the direction of an increase in entropy of the universe (system + surroundings). For a reversible process, a system at equilibrium, $\Delta S_{universe} = 0$. We can state this as:

> **BTW**
>
> *You should refer to Chapter 6 for a discussion of the terms* **system, surroundings,** *and* **universe**.

$$\Delta S_{universe} = \Delta S_{system} + \Delta S_{surroundings} > 0$$

for a spontaneous process

According to this second law, the entropy of the universe is continually increasing. The **third law of thermodynamics** states that for a pure crystalline substance at 0 K the entropy is zero.

Entropy

The qualitative entropy change (increase or decrease of entropy) for a system can sometimes be determined using a few simple rules:

- Entropy increases when the number of molecules increases during a reaction.

- Entropy increases with an increase in temperature.

- Entropy increases when a gas forms from either a liquid or solid.

- Entropy increases when a liquid forms from a solid.

In much the same fashion as the $\Delta H°$ was tabulated, the standard molar entropies ($S°$) of elements and compounds are tabulated. This is the entropy associated with 1 mole of a substance in its standard state. Unlike the enthalpies, the entropies of elements are not zero. For a reaction, it is possible to calculate the standard entropy change in the same fashion as the enthalpies of reaction:

$$\Delta S° = \Sigma\, \Delta S°_{products} - \Sigma\, \Delta S°_{reactants}$$

Gibbs Free Energy

One of the goals of chemists is to be able to predict if a reaction will be spontaneous. A reaction may be spontaneous if its ΔH is negative or if its ΔS is positive, but neither one is a reliable predictor by itself about whether a reaction will be spontaneous. Temperature also plays a part. A thermodynamic factor that considers the entropy, enthalpy, and temperature of the reaction would be the best indicator of spontaneity. This factor is the Gibbs free energy.

The Gibbs free energy (G) is:

$$G = H - TS$$

where T is the Kelvin temperature

Like most thermodynamic functions, it is only possible to measure the change in Gibbs free energy, so the relationship becomes:

$$\Delta G = \Delta H - T\Delta S$$

If there is a ΔG associated with a reaction and we reverse that reaction, the sign of the ΔG changes.

ΔG is the best indicator chemists have as to whether a reaction is spontaneous:

$\Delta G > 0 \rightarrow$ the reaction is not spontaneous;
energy must be supplied to cause the reaction to occur

$\Delta G < 0 \rightarrow$ the reaction is spontaneous

$\Delta G = 0 \rightarrow$ the reaction is at equilibrium

Just like with the enthalpy and entropy, the standard Gibbs free energy change, ($\Delta G°$), is calculated:

$$\Delta G° = \Sigma \, \Delta G°_{products} - \Sigma \, \Delta G°_{reactants}$$

The $\Delta G_f°$ of an element in its standard state is zero.

Free Energy and Reactions

We may also calculate $\Delta G°$ for a reaction by using the standard enthalpy and standard entropy of reaction:

$$\Delta G° = \Delta H°_{rxn} - T\Delta S°_{rxn}$$

It is possible to use this equation when the temperature is not standard. The $\Delta H°_{rxn}$ and the $\Delta S°_{rxn}$ values vary slightly with temperature. This slight variation allows an approximation of $\Delta G°$ at temperatures that are not standard.

In other cases, where the conditions are not standard, you should use the relationship:

$$\Delta G = \Delta G° + RT \ln Q$$

The ΔG symbol refers to the nonstandard Gibbs free energy value, $\Delta G°$ is the standard value, R is the gas constant (8.314 J/mol·K), T is the temperature (K), and Q is the reaction quotient first seen in Chapter 14. At equilibrium, this equation becomes:

$$0 = \Delta G° + RT \ln K \qquad \text{or} \qquad \Delta G° = -RT \ln K$$

> **EASY MISTAKE**
>
> The value of R uses joules and the value of the Gibbs free energy is typically in kilojoules. You will often need to change joules to kilojoules or kilojoules to joules.

Working Thermodynamic Problems

Let's work a few examples of problems involving enthalpy, entropy, and Gibbs free energy. We'll start off by working an entropy problem.

EXAMPLE

▶ Determine the value of $\Delta S°$ for each of the following reactions:

a. $H_2(g) + 1/2\ O_2(g) \rightarrow H_2O(g)$

b. $H_2(g) + 1/2\ O_2(g) \rightarrow H_2O(l)$

▶ To begin the problem, we need the standard entropy values for each of the reactants and products. These are the values we'll be using to complete this problem:

$H_2(g)$	131.0 J/mol · K
$O_2(g)$	205.0 J/mol · K
$H_2O(g)$	188.7 J/mol · K
$H_2O(l)$	69.9 J/mol · K

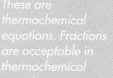

BTW

These are thermochemical equations. Fractions are acceptable in thermochemical equations.

▶ To finish this problem, we need to use the following relationship:

$$\Delta S° = \Sigma \, \Delta S°_{products} - \Sigma \, \Delta S°_{reactants}$$

▶ For part a:

$$\Delta S° = [(1 \text{ mol } H_2O) \, (188.7 \text{ J/mol} \cdot K)]$$

$$- [(1 \text{ mol } H_2) \, (131.0 \text{ J/mol} \cdot K) + (1/2 \text{ mol } O_2) \, (205.0 \text{ J/mol} \cdot K)]$$

$$\Delta S° = -44.8 \text{ J/ K}$$

▶ For part b:

$$\Delta S° = [(1 \text{ mol } H_2O) \, (69.9 \text{ J/mol} \cdot K)]$$

$$- [(1 \text{ mol } H_2) \, (131.0 \text{ J/mol} \cdot K) + (1/2 \text{ mol } O_2) \, (205.0 \text{ J/mol} \cdot K)]$$

$$\Delta S° = -163.6 \text{ J/K}$$

> **EASY MISTAKE**
> Make sure you use the correct value from the table of values. As seen in this example, the value for liquid water is not the same as for gaseous water.

> **BTW**
> Unlike heats of formation and Gibbs free energies of formation, the standard entropy values for elements are not zero.

> **EASY MISTAKE**
> Many people make the mistake of subtracting only the first of the reactant values from the product values. You must subtract the **sum** of the reactant values from the product values.

Now, let's concentrate on Gibbs free energy. Remember that it is our best indicator about whether a reaction is spontaneous.

EXAMPLE

▶ Determine the value of $\Delta G°$ for the following reaction:

$$2 \, NH_4Cl(s) + CaO(s) \rightarrow CaCl_2(s) + H_2O(l) + 2 \, NH_3(g)$$

▶ To begin the problem, we need the standard Gibbs free energy values for each of the reactants and products. These are the values we'll be using to complete this problem:

$$NH_4Cl(s) \quad -203.9 \text{ kJ/mol}$$

CaO(s)   −604.2 kJ/mol

CaCl$_2$(s)   −750.2 kJ/mol

H$_2$O(l)   −237.2 kJ/mol

NH$_3$(g)   −16.6 kJ/mol

▶ To finish this problem, we need to use the following relationship:

$$\Delta G° = \Sigma \, \Delta G°_{products} - \Sigma \, \Delta G°_{reactants}$$

$\Delta G° = [(1 \text{ mol CaCl}_2) \, (−750.2 \text{ kJ/mol}) + (1 \text{ mol H}_2\text{O})$
$(−237.2 \text{ kJ/mol}) + (2 \text{ mol NH}_3) \, (−16.6 \text{ kJ/mol})]$
$- [(1 \text{ mol CaO}) \, (−604.2 \text{ kJ/mol}) + (2 \text{ mol NH}_4\text{Cl}) \, (−203.9 \text{ kJ/mol})]$

$\Delta G° = −8.6 \text{ kJ}$

BTW

All sources may not contain identical values for thermodynamic values. Thermodynamic values are revised from time to time. You should use whatever values your reference book or the internet provides.

Here is another $\Delta G°$ problem.

▶ Using the relationship: $\Delta G° = \Delta H°_{rxn} - T\Delta S°_{rxn}$, calculate $\Delta G°$ for the following reaction:

$$2 \text{ Al}(s) + 3 \text{ S}(s) + 6 \text{ O}_2(g) \rightarrow \text{Al}_2(\text{SO}_4)_3(s)$$

▶ To begin the problem, we need the standard enthalpy and entropy values for each of the reactants and products. These are the values we'll be using to complete this problem:

	$\Delta H_f°$(kJ/mol)	$S°$(J/mol K)
Al(s)	0	28.32
Al$_2$(SO$_4$)$_3$(s)	−3441	239
O$_2$(g)	0	205.0
S(s)	0	31.88

► We need to use the following relationships before using the designated equation:

$$\Delta H° = \Sigma \, \Delta H°_{products} - \Sigma \, \Delta H°_{reactants}$$

$$\Delta S° = \Sigma \, \Delta S°_{products} - \Sigma \, \Delta S°_{reactants}$$

The $\Delta H_f°$ (and the $\Delta G_f°$) for elements in their standard states are zero. It is not necessary to search for these values. If you are in doubt, you can look for the values in a table.

► Solving these two relationships gives:

$$\Delta H° = [(1 \text{ mol } Al_2(SO_4)_3) \, (-3441 \text{ kJ/mol})] -$$
$$[(2 \text{ mol } Al) \, (0 \text{ kJ/mol}) + (3 \text{ mol } S) \, (0 \text{ kJ/mol}) +$$
$$(6 \text{ mol } O_2) \, (0 \text{ J/mol} \cdot K)]$$

$$\Delta H° = -3441 \text{ kJ}$$

$$\Delta S° = [(1 \text{ mol } Al_2(SO_4)_3) \, (239 \text{ J/mol} \cdot K)] - [(2 \text{ mol } Al)$$
$$(28.32 \text{ J/mol} \cdot K) + (3 \text{ mol } S) \, (31.88 \text{ J/mol} \cdot K) + (6 \text{ mol } O_2)$$
$$(205.0 \text{ J/mol} \cdot K)]$$

$$\Delta S° = -1143.28 \text{ J/K (unrounded)}$$

► We can now enter these results into the given equation. We will also need a kJ/J unit conversion.

$$\Delta G° = \Delta H°_{rxn} - T\Delta S°_{rxn}$$

$$\Delta G° = -3441 \text{ kJ}$$
$$\quad - (298 \text{ K}) \, (-1143.28 \text{ J/K}) \, (1 \text{ kJ}/10^3 \text{ J})$$

$$\Delta G° = -3100.30 = -3100. \text{ kJ}$$

The degree symbol indicates standard conditions. This means you should use 298 K as the temperature.

Now let's use some of these thermodynamic quantities to calculate the temperature at a phase change.

▶ At its melting point the heat of fusion for aluminum is 10.04 kJ/mol, and the entropy of fusion is 9.50 J/mol·K. Estimate the melting point of aluminum.

▶ The freezing point (or the melting point) is an equilibrium process. For any equilibrium process, ΔG is equal to 0. We can add this information to the following equation and rearrange:

$$\Delta G = \Delta H - T\Delta S = 0$$

$$T = \frac{\Delta H}{\Delta S}$$

▶ We can enter the values given in the problem, and a kJ/J conversion to find the temperature.

$$T = \frac{\Delta H}{\Delta S} = \frac{(10.04 \text{ kJ} / \text{mol})}{(9.50 \text{ J} / \text{mol} \cdot \text{K})} \frac{(10^3 \text{ J})}{(1 \text{ kJ})} = 1056.8 = 1.06 \times 10^3 \text{ K}$$

Another little twist: Let's use equilibrium data to calculate the Gibbs free energy of a reaction.

▶ The value of K_p for the following equilibrium at 298 K is 4.17×10^{14}. Calculate the value of $\Delta G°$, at this temperature for the equilibrium:

$$2 O_3(g) \rightleftarrows 3 O_2(g)$$

▶ The following equation relates the Gibbs free energy to an equilibrium constant:

$$\Delta G° = - RT \ln K$$

▶ We can enter the appropriate values into this equation to get the result:

$$\Delta G^\circ = -\left(\frac{8.314 \text{ J}}{\text{mol} \cdot \text{K}}\right)(298 \text{ K}) \ln (4.17 \times 10^{14})$$

$$\Delta G^\circ = -\left(\frac{8.314 \text{ J}}{\text{mol} \cdot \text{K}}\right)(298 \text{ K})(33.6641)$$

$$\Delta G^\circ = -83405.23 = -8.34 \times 10^4 \text{ J/mol}$$

Now let's go back the other way and use the Gibbs free energy to calculate the equilibrium constant for a reaction.

EXAMPLE

▶ The Gibbs free energy for the following process is 25.2 kJ/mol. Determine the equilibrium constant for this process at 25°C:

$$I_2(g) \leftrightharpoons 2 \text{ I}(g)$$

▶ The following equation relates the Gibbs free energy to an equilibrium constant:

$$\Delta G^\circ = -RT \ln K$$

▶ We will need to rearrange this equation:

$$\ln K = \frac{\Delta G^\circ}{-RT}$$

▶ We can now enter the given values into this relationship. We must not forget a kJ/J conversion and a °C/K conversion.

$$\ln K = \left(\frac{25.2 \text{ kJ / mol}}{-(8.314 \text{ J / mol} \cdot \text{K})(25 + 273)\text{K}}\right)\left(\frac{10^3 \text{ J}}{1 \text{ kJ}}\right)$$

$$\ln K = -10.1712 \text{ (unrounded)}$$

$$K = e^{-10.1712} = 3.8256 \times 10^{-5} = 3.82 \times 10^{-5}$$

Yet another twist: let's use concentrations to calculate a thermodynamic quantity.

EXAMPLE

A system, at 298 K, contains the gases NO, O_2, and NO_2. These gases are part of the following reaction:

$$2\ NO(g) + O_2(g) \rightarrow 2\ NO_2(g)$$

The concentrations of the gases are: NO = 2.00 M, O_2 = 0.500 M, and NO_2 = 1.00 M. Determine ΔG for the system.

Even though the system is at standard temperature, the system is not standard because the concentrations are not all one molar. For this reason, we must use the relationship:

$$\Delta G = \Delta G° + RT \ln Q$$

Before we can use this equation, we need to write the expression for Q:

$$Q = \frac{\left[NO_2\right]^2}{\left[NO\right]^2 \left[O_2\right]}$$

In addition to writing the Q expression, we also need to determine the value of $\Delta G°$ before using the equation. We can find $\Delta G°$ using the following relationship and values.

$$\Delta G° = \Sigma\ \Delta G°_{products} - \Sigma\ \Delta G°_{reactants}$$

NO(g)	86.71 kJ/mol
$O_2(g)$	0 kJ/mol (exactly)
$NO_2(g)$	51.84 kJ/mol

BTW

Anytime one or more of the following conditions are not met, the system is not standard, and this equation must be used. The conditions are 298 K, 1 M (for any reactant or product), and 1 atm (for any reactant or product that is a gas).

▶ Entering the Gibbs free energy values into the relationship gives:

$$\Delta G° = [(2 \text{ mol } NO_2) (51.84 \text{ kJ/mol})] - [(2 \text{ mol } NO) (86.71 \text{ kJ/mol}) + (1 \text{ mol } O_2) (0.000 \text{ kJ/mol})]$$

$$\Delta G° = -69.74 \text{ kJ}$$

▶ Returning to our key relationship for this problem:

$$\Delta G = \Delta G° + RT \ln Q$$

$$\Delta G = \Delta G° + RT \ln \frac{[NO_2]^2}{[NO]^2 [O_2]}$$

▶ We can now enter the values into the appropriate places. We also need a kJ/J conversion.

$$\Delta G = -69.74 \text{ kJ} + (8.314 \text{ J/mol} \cdot \text{K}) (298 \text{ K}) \left(\frac{1 \text{ kJ}}{10^3 \text{ J}} \right)$$

$$\ln \frac{[1.00]^2}{[2.00]^2 [0.500]}$$

$$\Delta G = -69.74 \text{ kJ} + (2.477572 \text{ kJ}) \ln \frac{[1.00]^2}{[2.00]^2 [0.500]}$$

$$\Delta G = -69.74 \text{ kJ} + (2.477572 \text{ kJ}) \ln 0.500$$

$$\Delta G = -69.74 \text{ kJ} + (2.477572 \text{ kJ}) (-0.693147)$$

$$\Delta G = -71.45732 = -71.46 \text{ kJ}$$

EXERCISES

EXERCISE 17-1

Answer these questions about entropy and free energy.

1. State the first, second, and third laws of thermodynamics.

2. Which of the following involves an increase in entropy?
   a. a solid melting
   b. a liquid freezing
   c. a gas condensing
   d. two gases reacting to produce a solid product
   e. a gas changing to a solid (undergoing deposition)

3. The Gibbs free energy for a spontaneous process has a _____ sign.

4. What is the mathematical definition of Gibbs free energy?

5. How can a process where ΔS_{system} is negative ever be spontaneous?

6. What are standard conditions?

7. What is the mathematical relationship necessary to adjust the Gibbs free energy to nonstandard conditions?

8. What is the mathematical relationship that relates the Gibbs free energy to the equilibrium constant?

9. What is the value of R used in thermochemical relationships?

10. How are K and Q similar? How are K and Q different?

11. What do the ΔH_f° and ΔG_f° values for elements have in common?

12. Determine the change in the standard entropy for the following reaction:

$$Ca(s) + S(s) + 3\ O_2(g) + 2\ H_2(g) \rightarrow CaSO_4{\cdot}2H_2O(s)$$

The $S°$ values are $Ca(s)$ = 41.63 J/mol·K, $S(s)$ = 31.88 J/mol·K, $O_2(g)$ = 205.0 J/mol·K, $H_2(g)$ = 131.0 J/mol·K, $CaSO_4{\cdot}2H_2O(s)$ = 194.0 J/mol·K

13. Determine the change in the standard Gibbs free energy for the following reaction:

$$Ca(s) + 2\ H_2SO_4(l) \rightarrow CaSO_4(s) + SO_2(g) + 2\ H_2O(l)$$

The $\Delta G_f°$ values are $Ca(s)$ = 0.00 kJ/mol, $H_2SO_4(l)$ = –689.9 kJ/mol, $SO_2(g)$ = –300.4 kJ/mol, $H_2O(l)$ = –237.2 kJ/mol, $CaSO_4(s)$ = –1320.3 kJ/mol

14. Determine the boiling point of acetone given that the heat of vaporization is 31.9 kJ/mol and the entropy of vaporization is 96.8 J/mol·K.

15. Estimate $\Delta G°$ for the following equilibrium:

$$Ag_2CrO_4(s) \leftrightarrows 2\ Ag^+(aq) + CrO_4{}^{2-}(aq) \quad K = 1.9 \times 10^{-16}$$

16. For the following reaction:

$$C(s) + CO_2(g) \rightarrow 2\ CO(g) \quad \Delta G° = 120.0\ kJ$$

if 25.00 g of C, 2.50 atm of CO, and 1.50 atm of CO_2 are placed in a 5.00 L container, calculate ΔG at 25°C.

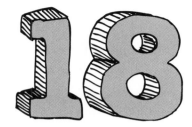

Electrochemistry

MUST KNOW

⚡ A redox reaction is a simultaneous reduction (gain of electrons) and oxidation (loss of electrons).

⚡ Balancing redox reactions involves balancing all the atoms in addition to the number of electrons lost and gained.

⚡ Galvanic (voltaic) cells use a redox reaction to produce electricity, while electrolytic cells use electricity to produce a desired redox reaction.

n this chapter we're going to learn how to balance redox equations, know the different types of electrochemical cells, and solve electrolysis problems. Have your computer or smartphone handy—you may need to find some information in electrochemical tables. We will be using the mole concept, so if you need some review, refer to Chapter 3, especially the mass/mole relationships. You might also need to review the section concerning net ionic equations in Chapter 4. Let's get started.

Redox Reactions

Electrochemical reactions involve redox reactions. **Redox** is a term that stands for reduction and oxidation. **Reduction** is the gain of electrons and **oxidation** is the loss of electrons. For example, if you place a piece of zinc metal in a solution containing the Cu^{2+} ion, some reddish solid forms on the surface of the zinc metal. That substance is copper metal. At the molecular level, the zinc metal is losing electrons to form the Zn^{2+} cation and the Cu^{2+} ion is gaining electrons to form copper metal. These two processes (called half-reactions) are:

$$Zn(s) \rightarrow Zn^{2+}(aq) + 2\ e^- \quad \text{(oxidation)}$$
$$Cu^{2+}(aq) + 2\ e^- \rightarrow Cu(s) \quad \text{(reduction)}$$

The electrons lost by the zinc metal are the same electrons gained by the copper(II) cation. The zinc metal is oxidized (loses electrons and increases its oxidation number) and the copper(II) ion is reduced (gains electrons and decreases its oxidation number).

The reactant causing the oxidation to take place is the **oxidizing agent** (the reactant undergoing reduction). In the preceding example, the oxidizing agent is the Cu^{2+} ion. The reactant undergoing oxidation is the **reducing agent**, because it furnishes the electrons that are necessary for the reduction half-reaction. Zinc metal is the reducing agent. The two half-reactions, oxidation and reduction, can be added together to give you the overall redox reaction. When doing this, the electrons must cancel—that is, there must be the same number of electrons lost as electrons gained:

$$Zn + Cu^{2+}(aq) + 2\ e^- \rightarrow Zn^{2+}(aq) + 2\ e^- + Cu$$

or

$$Zn + Cu^{2+}(aq) \rightarrow Zn^{2+}(aq) + Cu$$

Some redox reactions may be simply balanced by inspection. However, many are complex and require the use of a systematic method. There are two methods commonly used to balance redox reactions: the oxidation number method and the ion-electron method.

To balance a redox reaction using the oxidation number method, use the following rules in the order given:

- Assign oxidation numbers to all elements in the reaction.

- Identify what undergoes oxidation and what undergoes reduction by the change in oxidation numbers.

- Calculate the number of electrons lost in oxidation and electrons gained in reduction.

- Multiply one or both numbers by factors to make electron loss equal to electron gain and then use them as balancing coefficients.

- Complete the balancing by inspection.

- Check to see if the final equation is balanced.

To balance a redox reaction using the ion-electron method, use the following rules in the order given: (Several variations of this method are used.)

- Assign oxidation numbers and begin the half-reactions, one for oxidation and one for reduction.

- Balance all atoms except O and H.

- Balance oxygen atoms.

- Balance hydrogen atoms.

■ Balance charges by adding electrons.

■ Adjust the half-reactions to give equal numbers of electrons.

■ Add and cancel (the electrons must cancel).

■ Check to see if the final equation is balanced.

EXTRA HELP

Let's tackle balancing the reaction below using the ion-electron method:

$$Cu(s) + HNO_3(aq) \rightarrow Cu(NO_3)_2(aq) + NO(g) + H_2O(l)$$

Follow these steps:

1. Convert the unbalanced redox reaction to the ionic form.

   In this reaction, we show the nitric acid in the ionic form because it's a strong acid. Copper(II) nitrate is soluble (indicated by (*aq*)), so it's shown in its ionic form. Because NO(*g*) and water are molecular compounds, they remain shown in the molecular form:

   $$Cu(s) + H^+ + NO_3^- \rightarrow Cu^{2+} + 2\,NO_3^- + NO(g) + H_2O(l)$$

2. If necessary, assign oxidation numbers and then write two half-reactions (oxidation and reduction) showing the chemical species that have had their oxidation numbers changed.

   In some cases, it's easy to tell what has been oxidized and reduced; but in other cases, it isn't as easy. Start by going through the example reaction and assigning oxidation numbers. You can then use the chemical species that have had their oxidation numbers changed to write your unbalanced half-reactions:

   $$Cu(s) + H^+ + NO_3^- \rightarrow Cu^{2+} + 2\,NO_3^- + NO(g) + H_2O(l)$$

Cu(s)	H⁺	NO₃⁻	Cu²⁺	2 NO₃⁻	NO(g)	H₂O(l)
0	+1	+5(–2)3	+2	+5(–2)3	+2 –2	(+1)2 –2

Look closely. Copper changed its oxidation number (from 0 to 2) and so has nitrogen (from +5 to +2). Your unbalanced half-reactions are

$$Cu(s) \rightarrow Cu^{2+}$$
$$NO_3^- \rightarrow NO$$

3. Balance all atoms, with the exception of oxygen and hydrogen.

   It is a good idea to wait until the end to balance hydrogen and oxygen atoms, so always balance the other atoms first. However, in this particular case, both the copper and nitrogen atoms already balance with one each on both sides:

$$Cu(s) \rightarrow Cu^{2+}$$
$$NO_3^- \rightarrow NO$$

4. Balance the oxygen atoms.
   - In acid solutions, take the number of oxygen atoms needed and add that same number of water molecules to the side that needs oxygen.
   - In basic solutions, add 2 OH^- to the side that needs oxygen for every oxygen atom that is needed. Then, to the other side of the equation, add half as many water molecules as OH^- anions used.

   An acidic solution will have some acid or H^+ shown; a basic solution will have an OH^- present. The example equation is in acidic conditions (nitric acid, HNO_3, which in ionic form is $H^+ + NO_3^-$). There's nothing to do on the half-reaction involving the copper because there are no oxygen atoms present. But you do need to balance the oxygen atoms in the second half-reaction:

$$Cu(s) \rightarrow Cu^{2+}$$
$$NO_3^- \rightarrow NO + 2\,H_2O$$

5. Balance the hydrogen atoms.
   - In acid solutions, take the number of hydrogen atoms needed and add that same number of H^+ cations to the side that needs hydrogen.
   - In basic solutions, add one water molecule to the side that needs hydrogen for every hydrogen atom that's needed. Then, to the

other side of the equation, add as many OH⁻ anions as water molecules used.

The example equation is in acidic conditions. You need to balance the hydrogen atoms in the second half-reaction:

$$Cu(s) \rightarrow Cu^{2+}$$
$$4\,H^+ + NO_3^- \rightarrow NO + 2\,H_2O$$

6. Balance the ionic charge on each half-reaction by adding electrons.

$$Cu(s) \rightarrow Cu^{2+} + 2\,e^- \text{ (oxidation)}$$
$$3\,e^- + 4\,H^+ + NO_3^- \rightarrow NO + 2\,H_2O \text{ (reduction)}$$

7. Balance electron loss with electron gain between the two half-reactions.

The electrons that are lost in the oxidation half-reaction are the same electrons that are gained in the reduction half-reaction. The number of electrons lost and gained must be the same. But Step 6 shows a loss of 2 electrons and a gain of 3. So, you must adjust the numbers using appropriate multipliers for both half-reactions. In this case, you have to find the lowest common denominator between 2 and 3. It's 6, so multiply the first half-reaction by 3 and the second half-reaction by 2.

EASY MISTAKE The electrons should end up on opposite sides of the equation in the two half-reactions. Also, remember that you're using ionic charge, not oxidation numbers.

$$3 \times [Cu(s) \rightarrow Cu^{2+} + 2\,e^-] = 3\,Cu(s) \rightarrow 3\,Cu^{2+} + 6\,e^-$$
$$2 \times [3\,e^- + 4\,H^+ + NO_3^- \rightarrow NO + 2\,H_2O] = 6\,e^- + 8\,H^+ + 2\,NO_3^-$$
$$\rightarrow 2\,NO + 4\,H_2O$$

8. Add the two half-reactions together, and cancel anything common to both sides. The electrons should always cancel.

$$3\,Cu + \cancel{6\,e^-} + 8\,H^+ + 2\,NO_3^- \rightarrow 3\,Cu^{2+} + \cancel{6\,e^-} + 2\,NO + 4\,H_2O$$

9. Convert the equation back to the molecular form by adding the spectator ions.

If it is necessary to add spectator ions to one side of the equation, add the same number to the other side of the equation. For example, there are 8 H$^+$ on the left side of the equation. In the original equation, the H$^+$ was in the molecular form of HNO$_3$ so you will need to add the NO$_3^-$ spectator ions back in. You already have 2 on the left, so simply add 6 more. Then add 6 NO$_3^-$ to the right-hand side to keep things balanced. Those will be the spectator ions that you need for the Cu^{2+} cation to convert it back to the molecular form that you want.

$$3 \, Cu(s) + 8 \, HNO_3(aq) \rightarrow 3 \, Cu(NO_3)_2(aq) + 2 \, NO(g) + 4 \, H_2O(l)$$

10. Check to make sure that all the atoms are balanced and all the charges are balanced (if working with an ionic equation at the beginning).

Reactions that take place in base are just as easy, as long as you follow the rules.

Galvanic (Voltaic) Cells

Galvanic (voltaic) cells produce electricity by using a redox reaction. Let's take that zinc/copper redox reaction that we studied before (the direct electron transfer one) and make it a galvanic cell by separating the oxidation and reduction half-reactions.

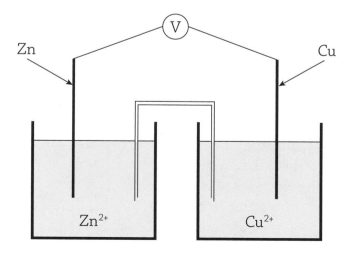

Instead of one container, we will use two. We place a piece of zinc metal in one and a piece of copper metal in another. We add a solution of aqueous zinc sulfate (or some other water soluble zinc salt) to the beaker containing the zinc electrode and an aqueous solution of copper(II) sulfate (or some other water soluble copper(II) salt) to the beaker containing the copper metal. The zinc and copper metals will form the **electrodes** of the cell, the solid portion of the cell that conducts the electrons that are involved in the redox reaction. The solutions in which we immerse the electrodes are the electrode compartments.

We connect the electrodes by a wire and complete the circuit with a salt bridge. A **salt bridge** is often an inverted U-tube containing a concentrated electrolyte solution, such as KNO_3. The anions in the salt bridge will migrate through the solution into the beaker containing the zinc metal, and the salt-bridge cations will migrate in the opposite direction. In this way, both electrode compartments maintain electrical neutrality. The zinc electrode is being oxidized in one beaker, and the copper(II) ions in the other beaker are being reduced to copper metal. The same redox reaction is happening in this indirect electron transfer as happened in the direct one:

$$Zn(s) + Cu^{2+}(aq) \rightarrow Zn^{2+}(aq) + Cu(s)$$

The difference is that the electrons are now flowing through a wire from the oxidation half-reaction to the reduction half-reaction. The flow of electrons through a wire is electricity. If we connect a voltmeter to the wire connecting the two electrodes, we would measure a current of 1.10 V. This galvanic cell is a Daniell cell.

The electrode at which oxidation is taking place is called the **anode** and the electrolyte solution in which it is immersed is called the **anode compartment**. The electrode at which reduction takes place is the **cathode** and its solution is the **cathode compartment**. We label the anode with a negative sign (–), while the cathode has a positive sign (+). The electrons flow from the anode to the cathode (negative to positive).

Don't forget!
Oxidation is an
anode process.

Sometimes the half-reaction(s) involved in the cell do not have a solid conductive part to act as the electrode, so an inert (inactive) electrode, a solid conducting electrode that does not take part in the redox reaction, is used. Graphite and platinum are common inert electrodes.

Cell notation is a shorthand notation of representing a galvanic cell. To write the cell notation for the Daniell cell you:

Write the chemical formula of the anode.	$Zn(s)$
Draw a single vertical line to represent the phase boundary between the solid anode and the anode compartment.	$Zn(s) \mid$
Write the reactive part of the anode compartment with its initial concentration (if known) in parentheses (assume 1 M in this case).	$Zn(s) \mid Zn^{2+}(1\ M)$
Draw a double vertical line to represent the salt bridge connecting the two electrode compartments.	$Zn(s) \mid Zn^{2+}(1\ M) \mid\mid$
Write the reactive part of the cathode compartment with its initial concentration (if known) shown in parentheses.	$Zn(s) \mid Zn^{2+}(1\ M) \mid\mid Cu^{2+}(1\ M)$
Draw a single vertical line representing the phase boundary between the cathode compartment and the cathode.	$Zn(s) \mid Zn^{2+}(1\ M) \mid\mid Cu^{2+}(1\ M) \mid$
And, finally, write the chemical formula of the cathode.	$Zn(s) \mid Zn^{2+}(1\ M) \mid\mid Cu^{2+}(1\ M) \mid Cu(s)$

If there is an inert electrode present, then show where the inert electrode is with its phase boundary. If the electrode components are in the same phase, then separate them by commas; if not, use a vertical phase boundary line. For example, consider the following redox reaction:

$$Ag^+(aq) + Fe^{2+}(aq) \rightarrow Fe^{3+}(aq) + Ag(s)$$

The oxidation of the iron(II) ion to iron(III) doesn't involve a solid, so we must use an inert electrode, such as platinum. The cell notation would then be:

$$Pt(s) \mid Fe^{2+}(aq), Fe^{3+}(aq) \mid\mid Ag^+(aq) \mid Ag(s)$$

Standard Reduction Potentials ($E°$)

In the discussion of the Daniell cell, we indicated that this cell produces a voltage of 1.10 V. This voltage is really the difference in potential between the two half-cells. The cell potential (really the half-cell potentials) is dependent on concentration and temperature, but initially we'll simply look at the half-cell potentials at the standard state of 298 K (25°C) and all components in their standard states (1 M concentration of all solutions, 1 atm pressure for any gases and pure solid electrodes). Half-cell potentials appear in tables as the reduction potentials, that is, the potentials associated with the reduction reaction. We define the hydrogen half-reaction (2 $H^+(aq)$ + 2e⁻ → $H_2(g)$) as the standard and it has been given a value of exactly 0.00 V. We measure all the other half-reactions relative to it; some are positive and some are negative. Let's look at a table of standard reduction potentials:

Standard Reduction Potentials (at 25°C)

Reduction Half-Reaction		$E°(V)$
$F_2(g) + 2e^-$	$\rightarrow 2F^-(aq)$	2.87
$Ce^{4+}(aq) + e^-$	$\rightarrow Ce^{3+}(aq)$	1.61
$MnO_4^-(aq) + 8H^+(aq) + 5e^-$	$\rightarrow Mn^{2+}(aq) + 4H_2O(l)$	1.51
$Cl(g) + 2e^-$	$\rightarrow 2Cl^-(aq)$	1.36
$O_2(g) + 4H^+(aq) + 4e^-$	$\rightarrow 2H_2O(l)$	1.23
$Br_2(l) + 2e^-$	$\rightarrow 2Br^-(aq)$	1.06
$NO_3^-(aq) + 4H^+(aq) + 3e^-$	$\rightarrow NO(g) + 2H_2O(l)$	0.96
$Ag^+(aq) + e^-$	$\rightarrow Ag(s)$	0.80
$Fe^{3+}(aq) + e^-$	$\rightarrow Fe^{2+}(aq)$	0.77
$I_2(s) + 2e^-$	$\rightarrow 2I^-(aq)$	0.54
$Cu^{2+}(aq) + 2e^-$	$\rightarrow Cu(s)$	0.34
$AgCl(s) + e^-$	$\rightarrow Ag(s) + Cl^-(aq)$	0.222
$Sn^{4+}(aq) + 2e^-$	$\rightarrow Sn^{2+}(aq)$	0.15
$2H^+(aq) + 2e^-$	$\rightarrow H_2(g)$	0.000
$Pb^{2+}(aq) + 2e^-$	$\rightarrow Pb(s)$	-0.126
$Ni^{2+}(aq) + 2e^-$	$\rightarrow Ni(s)$	-0.25
$Cr^{3+}(aq) + e^-$	$\rightarrow Cr^{2+}(aq)$	-0.41
$Fe^{2+}(aq) + 2e^-$	$\rightarrow Fe(s)$	-0.44
$Zn^{2+}(aq) + 2e^-$	$\rightarrow Zn(s)$	-0.76
$Ba^{2+}(aq) + 2e^-$	$\rightarrow Ba(s)$	-1.57
$Al^{3+}(aq) + 3e^-$	$\rightarrow Al(s)$	-1.66
$Mg^{2+}(aq) + 2e^-$	$\rightarrow Mg(s)$	-2.37
$Na^+(aq) + e^-$	$\rightarrow Na(s)$	-2.714
$Li^+(aq) + e^-$	$\rightarrow Li(s)$	-3.045

We can use a table of standard reduction potentials, such as the one on the preceding page, to write the overall cell reaction and to calculate the standard cell potential, the potential (voltage) associated with the cell at standard conditions. There are a couple of things to remember when using these standard reduction potentials to generate the cell reaction and cell potential:

■ Since the standard cell potential is for a galvanic cell, it must be a positive value, $E° > 0$.

■ Since one half-reaction must involve oxidation, we must reverse one of the half-reactions shown in the table of reduction potentials to indicate the oxidation. If we reverse the half-reaction, we must also reverse the sign of the standard reduction potential.

■ Since oxidation occurs at the anode and reduction at the cathode, the standard cell potential can be calculated from the standard reduction potentials of the two half-reactions involved in the overall reaction by using the equation:

$$E°_{cell} = E°_{cathode} - E°_{anode} > 0$$

But remember both the $E°_{cathode}$ and $E°_{anode}$ values are shown as reduction potentials, used directly from the table without reversing.

Once you have calculated the standard cell potential, then the reaction can be written by reversing the half-reaction associated with the anode (show it as oxidation) and adding the two half-reactions.

Don't forget that the number of electrons lost must equal the number of electrons gained. If they are not equal, use appropriate multipliers to ensure that they are equal.

EXAMPLE

▶ Calculate the potential of a galvanic cell using the following half-cells:

$$Ni^{2+} + 2\ e^- \rightarrow Ni(s) \qquad E° = -0.25\ V$$

$$Ag^+ + e^- \rightarrow Ag(s) \qquad E° = 0.80\ V$$

▶ First, calculate the cell potential using:

$$E°_{cell} = E°_{cathode} - E°_{anode} > 0$$

▶ Since the cell potential must be positive (a galvanic cell), there is only one arrangement of −0.25 V and 0.80 V that can result in a positive value:

$$E°_{cell} = 0.80\ V - (-0.25\ V) = 1.05\ V$$

▶ This means that the Ni electrode is the anode and is involved in oxidation. Therefore, we reverse the reduction half-reaction involving Ni, changing the sign of the standard half-cell potential, and add it to the silver half-reaction. We must multiply the silver half-reaction by 2 to equalize electron loss and gain, but the half-cell potential is not altered:

$$Ni(s) \rightarrow Ni^{2+}(aq) + 2\ e^- \qquad E° = 0.25\ V$$

$$2(Ag^+(aq) + e^- \rightarrow Ag(s)) \qquad E° = 0.80\ V$$

$$Ni(s) + 2\ Ag^+(aq) \rightarrow Ni^{2+}(aq) + Ag(s) \qquad E°_{cell} = 1.05\ V$$

IRL Reactions such as these are used to produce electricity for a wide variety of purposes—starting your car, powering your phone, and so on. Be careful to distinguish between a cell and a battery. A cell is a single redox reaction, while a battery is several cells connected together. It is proper to talk about an automobile battery since it is six cells, each producing a little over 2 volts, to give you 12 volts (actually a little more). It is incorrect to talk about a flashlight battery since there is only one cell. You should say a flashlight cell.

Nernst Equation

Thus far, we have based all our calculations on the standard cell potential or standard half-cell potentials—that is, standard state conditions. However, many times the cell is not at standard conditions—commonly the concentrations are not 1 M. We may calculate the actual cell potential, E_{cell}, by using the **Nernst equation**:

$$E_{cell} = E°_{cell} - \left(\frac{RT}{nF}\right)\ln Q = E°_{cell} - \left(\frac{0.0592}{n}\right)\log Q \text{ at } 25°C$$

R is the ideal gas constant, T is the Kelvin temperature, n is the number of electrons transferred, F is Faraday's constant (96,485 J/V or 96,485 coulombs/mol), and Q is the reaction quotient. The second form, involving the log Q, is the more useful form. If you know the cell reaction, the concentrations of ions, and the $E°_{cell}$, then you can calculate the actual cell potential. Another useful application of the Nernst equation is in the calculation of the concentration of one of the reactants from cell potential measurements. Knowing the actual cell potential and the $E°_{cell}$ allows you to calculate Q, the reaction quotient. Knowing Q and all but one of the concentrations allows you to calculate the unknown concentration. Another application of the Nernst equation is concentration cells. A **concentration cell** is an electrochemical cell in which the same chemical species are used in both cell compartments but differing in concentration. Because the half-reactions are the same, the $E°_{cell} = 0.00$ V. Then simply substituting the appropriate concentrations into the activity quotient allows calculation of the actual cell potential.

When using the Nernst equation on a cell reaction in which the overall reaction is not supplied, only the half-reactions and concentrations, there are two equivalent methods to work the problem. The first way is to write the overall redox reaction based on $E°$ values and then apply the Nernst equation. If the E_{cell} turns out to be negative, it indicates that the reaction is not a spontaneous one (an electrolytic cell) or that the reaction is written backward if it is supposed to be a galvanic cell. If it is supposed to be a galvanic cell, then all you need to do is reverse the overall reaction and change the sign on the E_{cell} to positive. The other method involves using the Nernst equation with the individual half-reactions and then combining them depending on whether it is a galvanic cell. The only disadvantage to the second method is that you must use the Nernst equation twice. Either method should lead you to the correct answer.

EXAMPLE

▶ Calculate the potential of the following half-cell:

$$Cr_2O_7^{2-}(aq) + 14\ H^+(aq) + 6\ e^- \rightarrow 2\ Cr^{3+}(aq) + 7\ H_2O(l)\ E° = 1.33\ V$$

▶ We are given that it contains 0.10 M $K_2Cr_2O_7$, 0.20 M $Cr^{3+}(aq)$, and 1.0 × 10^{-4} M $H^+(aq)$.

▶ The solution is:

$$E = E°_{cell} - \left(\frac{0.0592}{n}\right)\log\frac{[Cr^{3+}]^2}{[Cr_2O_7^{2-}][H^+]^{14}}\ \text{ignoring } H_2O$$

(as always)

$$E = E°_{cell} - \left(\frac{0.0592}{6}\right)\log\frac{[0.20]^2}{[0.10][1\times10^{-4}]^{14}} = 0.78\ V$$

Electrolytic Cells

Electrolytic cells use electricity from an external source to produce a desired redox reaction. Electroplating and the recharging of an automobile battery are examples of electrolytic cells.

In the operation of both galvanic and electrolytic cells, there is a reaction occurring on the surface of each electrode. For example, the following reaction takes place at the cathode of a cell:

$$Cu^{2+}(aq) + 2\ e^- \rightarrow Cu(s)$$

The rules of stoichiometry also apply in this case. In electrochemical cells, we must consider not only the stoichiometry related to chemical formulas, but also the stoichiometry related to electric currents. The half-reaction under consideration not only involves 1 mole of each of the copper species, but also 2 moles of electrons. We can construct a mole ratio that includes moles of electrons or we could construct a mole ratio using faradays. A **faraday (F)** is a mole of electrons. Thus, we could use either of the following ratios for the copper half-reaction:

$$\frac{1 \text{ mol Cu}^{2+}}{2 \text{ mol e}^-} = \frac{1 \text{ mol Cu}^{2+}}{2 \text{ F}}$$

The SI base unit for electric current is the **ampere (A)**. In addition to being an SI base unit, an ampere is a coulomb (C) per second, and a faraday is 96,485 C/mol of electrons. Therefore:

$$1 \text{ ampere} = 1 \text{ A} = \frac{1 \text{ C}}{1 \text{ s}} \quad \text{and} \quad 1 \text{ faraday} = 1 \text{ F} = \frac{96,485 \text{ C}}{1 \text{ mol electrons}}$$

EXAMPLE

▶ If liquid titanium(IV) chloride (acidified with HCl) is electrolyzed by a current of 1.000 A for 2 h, how many grams of titanium will be produced?

▶ First, write the half-reaction:

$$Ti^{4+}(l) + 4 \text{ e}^- \rightarrow Ti(s)$$

▶ The information from the problem and the half-reaction are:

$$Ti^{4+}(l) + 4 \text{ e}^- \qquad \rightarrow Ti(s)$$

2.000 h   ? grams

1.000 A

▶ To simplify the solution, we will write amperes as its definition of coulomb/second.

$$\text{grams Ti} = (2.000 \text{ h})\left(\frac{3600 \text{ s}}{1 \text{ h}}\right)\left(\frac{1.000 \text{ C}}{\text{s}}\right)\left(\frac{1 \text{ mol } e^-}{96485 \text{ C}}\right)\left(\frac{1 \text{ mol Ti}}{4 \text{ mol } e^-}\right)\left(\frac{47.87 \text{ g Ti}}{1 \text{ mol Ti}}\right)$$

$$= 0.89305073 = 0.8930 \text{ g Ti}$$

Electrolysis

Electrolysis is a process by which direct current is used to cause a nonspontaneous reaction to occur. (This is essentially an electrolytic cell.) Electrolysis is used to separate certain elements from naturally occurring ores.

One way to represent an electrolysis reaction is:

$$2 \text{ KF}(l) \xrightarrow{\text{electrolysis}} 2 \text{ K}(l) + \text{F}_2(g)$$

In this reaction the cathode and anode reactions are:

cathode $\qquad$ $\text{K}^+ + 1 \text{ e}^- \rightarrow \text{K}$

anode $\qquad$ $2 \text{ F}^- \rightarrow \text{F}_2 + 2 \text{ e}^-$

We are assuming the standard potentials for aqueous solution are approximately equal to the potentials in molten KF.

The cathode reaction comes directly from a table of standard reduction potentials, while the anode reaction is the reverse reaction from such a table. However, how did we know that it was the fluorine reaction requiring reversal?

The fluorine half-reaction in the table is:

$$\text{F}_2 + 2 \text{ e}^- \rightarrow 2 \text{ F}^-$$

EASY MISTAKE

Oxidation is an anode process, even in electrolysis, and reduction is always a cathode process.

If we examine the reactant, we find that the compound, KF, is an ionic compound containing potassium ions and fluoride ions. For this reason, we could replace the KF(l) in the original equation with $K^+(l) + F^-(l)$. These two ions, either alone or in combination, are the only substances, other than electrons, that can appear on the reactant side of the half-reactions. One of these ions, the fluoride ion, appears in the fluorine half-reaction. Since KF, and therefore F^-, is a reactant, we must reverse the fluorine half-reaction to place the fluoride ion on the reactant side. The original KF has no F_2, so F_2 cannot be a reactant.

What happens if we replace our reactant, KF(l), with KF(aq)? This apparently minor change makes a big difference in the results. The potassium ions and the fluoride ions are still present, so they are still under consideration, but we also need to consider water. Water appears in many places in a table of reduction potentials. We must examine every place it appears alone on a side or with one of the ions we know to be present, K^+ and F^-. The potassium and fluorine half-reactions, with their reduction potentials, are:

> **BTW**
>
> *If you reverse the reaction, you must reverse the sign of the cell potential.*

$$K^+(aq) + 1\ e^- \rightarrow K(s) \qquad E° = -2.93\ V$$
$$F_2(g) + 2\ e^- \rightarrow 2\ F^-(aq) \qquad E° = +2.87\ V$$

We need to reverse the fluorine half-reaction to place the fluoride ion on the reactant side:

$$2\ F^-(aq) \rightarrow F_2(g) + 2\ e^- \qquad E° = -2.87\ V$$

Water, the other potential reactant, appears in the following half-reactions:

$$2\ H_2O(l) + 2\ e^- \rightarrow H_2(g) + 2\ OH^-(aq) \qquad E° = -0.83\ V$$
$$4\ H^+(aq) + O_2(g) + 4\ e^- \rightarrow 2\ H_2O(l) \qquad E° = +1.23\ V$$

We need to reverse the second of these reactions to place the water on the reactant side.

$$2 H_2O(l) \rightarrow 4 H^+(aq) + O_2(g) + 4 e^- \qquad E° = -1.23 \text{ V}$$

The reactions that may occur at the cathode are:

$$K^+(aq) + 1 e^- \rightarrow K(s) \qquad E° = -2.93 \text{ V}$$

$$2 H_2O(l) + 2 e^- \rightarrow H_2(g) + 2 OH^-(aq) \qquad E° = -0.83 \text{ V}$$

The reactions that may occur at the anode are:

$$2 F^-(aq) \rightarrow F_2(g) + 2 e^- \qquad E° = -2.87 \text{ V}$$

$$2 H_2O(l) \rightarrow 4 H^+(aq) + O_2(g) + 4 e^- \qquad E° = -1.23 \text{ V}$$

We must narrow the options. There will be only one cathode reaction and only one anode reaction. How do we pick the correct half-reactions? If one of the half-reactions were spontaneous (positive), we would pick it for that electrode. (If more than one were spontaneous, we would pick the largest positive value.) All four half-reactions in this case are nonspontaneous (negative). This is typical for electrolysis, because you are using electrical energy to force a nonspontaneous process to take place.

We will begin with the cathode. The reaction that will occur will be the one requiring the less energy. This will be the less negative (higher) value. We can eliminate all other reduction half-reactions. Therefore, the cathode half-reaction must be:

> **BTW**
>
> *There is a phenomenon known as overvoltage, which leads to variations in these rules. The water/oxygen half-reaction, shown here, is often subject to this complication. If overvoltage is present, the oxygen value may become -1.40 V instead of -1.23 V.*

$$2 H_2O(l) + 2 e^- \rightarrow H_2(g) + 2 OH^-(aq) \qquad E° = -0.83 \text{ V}$$

The same rules apply to the anode. From these rules, we see that the anode half-reaction must be:

$$2 H_2O(l) \rightarrow 4 H^+(aq) + O_2(g) + 4 e^- \qquad E° = -1.23 \text{ V}$$

Once we know the two half-reactions that will occur, we can determine the cell reaction using the final steps for balancing redox equations.

$$2 (2 H_2O(l) + 2 e^- \rightarrow H_2(g) + 2 OH^-(aq))$$
$$2 H_2O(l) \rightarrow 4 H^+(aq) + O_2(g) + 4 e^-$$
$$\downarrow$$
$$6 H_2O(l) + 4 e^- \rightarrow 2 H_2(g) + 4 OH^-(aq) + 4 H^+(aq) + O_2(g) + 4 e^-$$

The 4 $H^+(aq)$ and the 4 $OH^-(aq)$ become 4 $H_2O(l)$:

$$6 H_2O(l) + 4 e^- \rightarrow 2 H_2(g) + 4 H_2O(l) + O_2(g) + 4 e^-$$

We can now cancel to get the final balanced equation:

$$2 H_2O(l) \rightarrow 2 H_2(g) + O_2(g)$$

In an electrolysis process, such as this one, the potassium ions and the fluoride ions are spectator ions. They must be present for the procedure to work, but they will remain unchanged.

EXERCISES

EXERCISE 18-1

Answer the following questions.

1. Define oxidation and reduction.

2. Assign oxidation numbers to each element in each of the following substances.
   - a. H_2O
   - b. Na_2SO_4
   - c. $FeCl_3$
   - d. K_2O_2
   - e. $CaCr_2O_7$

3. What is the sign of the anode in a galvanic cell? What is the sign of the anode in an electrolytic cell?

4. In a galvanic cell, reduction occurs at which electrode? In an electrolytic cell, reduction occurs at which electrode?

5. What happens to the oxidizing agent during a redox reaction?

6. What is a faraday?

7. Balance the following half-reactions.
   - a. $2\,e^- + Cl_2 \rightarrow$ _____
   - b. $3\,e^- + CrO_4^{2-} + 4\,H_2O \rightarrow$ _____ $+ 8\,OH^-$
   - c. _____ $+ FeO_4^{2-} + 8\,H^+ \rightarrow Fe^{3+} + 4\,H_2O$
   - d. $Pb^{4+} +$ _____ $\rightarrow Pb^{2+}$
   - e. $S_2^{2-} \rightarrow 2\,S +$ _____

8. Metallic magnesium can be made by the electrolysis of molten $MgCl_2$. What mass of Mg is formed by passing a current of 3.50 A through molten $MgCl_2$ for a period of 550 min?

9. Use standard electrode potentials to calculate $E°_{cell}$ for the disproportionation of copper(I) ion: $2\,Cu^+(aq) \rightarrow Cu(s) + Cu^{2+}(aq)$. The standard reduction potentials are:

$$Cu^+(aq) + 1\,e^- \rightarrow Cu(s) \qquad E° = +0.52\,V$$

$$Cu^{2+}(aq) + 1\,e^- \rightarrow Cu^+(s) \quad E° = +0.15\,V$$

10. Calculate E_{cell} for the reaction in question 9, if $[Cu^+] = 0.25$ M and $[Cu^{2+}] = 1.50$ M.

11. Use the reduction potentials given in this chapter and the following half-reaction:

$$I_2(s) + 2\,e^- \rightarrow 2\,I^-(aq) \qquad E° = +0.53\,V$$

Write the balanced chemical equation for the electrolysis of a potassium iodide, KI, solution.

EXERCISE 18-2

Balance the following redox equations.

1. $S^{2-}(aq) + Cr_2O_7^{2-}(aq) + H^+(aq) \rightarrow S(s) + Cr^{3+}(aq) + H_2O(l)$

2. $Cl_2(g) + H_2O(l) + SO_2(g) \rightarrow SO_4^{2-}(aq) + Cl^-(aq) + H^+(aq)$

3. $Mn^{2+}(aq) + PbO_2(s) + H^+(aq) \rightarrow MnO_4^-(aq) + Pb^{2+}(aq) + H_2O(l)$

4. $SnO_2^{2-}(aq) + Bi^{3+}(aq) + OH^-(aq) \rightarrow SnO_3^{2-}(aq) + Bi(s) + H_2O(l)$

5. $ClO_2^-(aq) + MnO_4^-(aq) \rightarrow MnO(s) + ClO_4^-(aq)$ (basic)

Flashcard App

19 Chemistry of the Elements

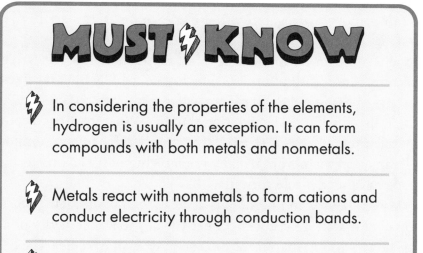

MUST KNOW

- In considering the properties of the elements, hydrogen is usually an exception. It can form compounds with both metals and nonmetals.

- Metals react with nonmetals to form cations and conduct electricity through conduction bands.

- Nonmetals form anions when reacting with metals.

In this chapter we're going to examine some properties of the elements, as well as the periodic trends that we can observe in these properties. You might want to review briefly the periodic trends we discussed in Chapter 8. We will also discuss coordination compounds and complex ions. Let's take a look.

Hydrogen

Hydrogen is the only nonmetal in Group 1A (1) on the periodic table. Some periodic tables move hydrogen to other places or list it in more than one place.

Hydrogen can form compounds with all elements except the noble gases. In compounds with nonmetals, hydrogen usually behaves like a metal instead of a nonmetal. Therefore, when hydrogen combines with a nonmetal, it usually has a +1-oxidation number. When hydrogen combines with a metal, it usually has a −1-oxidation number. Hydrogen compounds with the transition metals are usually nonstoichiometric. Nonstoichiometric compounds have no definite formula.

All Brønsted–Lowry and Arrhenius acids contain hydrogen. The formation or transfer of hydrogen ions is the key to the behavior of all Brønsted–Lowry and Arrhenius acids. You may wish to review the behavior of acids in Chapters 4 and 15.

The standard cell potential for the reduction of hydrogen ions to hydrogen gas is, by definition, 0.00 V. This potential is for the standard hydrogen electrode, SHE, which is the reference to which we compare all other cell potentials (see Chapter 18). All metals above hydrogen on the activity series will displace hydrogen gas from acids (see Chapter 4). Metals below hydrogen will not displace hydrogen gas.

General Properties of Metals

In compounds, metals are cations. That is, they will have a positive charge. The cations may be shown with the corresponding anions in solution, or in some problems, only the metal cation may be shown. In both cases, you may see designations such as Al^{3+}. The metals contained in the leftmost two columns on the periodic table form +1 and +2 ions, respectively. All other columns contain metals that may adopt more than one oxidation state. Variations in oxidation numbers are prevalent among the transition metals. The highest oxidation number known for any metal in a compound is +8.

Compounds containing metals must also contain a nonmetal or a polyatomic anion. There are no compounds found in a general chemistry course where a compound only contains metals.

Metals react with nonmetals. These reactions are oxidation-reduction reactions. (See Chapters 4 and 18.) Oxidation of the metal occurs in conjunction with reduction of the nonmetal. In most cases, only simple compounds will form. For example, oxygen, O_2, reacts with nearly all metals to form oxides (compounds containing O^{2-}). Exceptions are the reaction with sodium where sodium peroxide, Na_2O_2, forms and the reaction with potassium, rubidium, and cesium where the superoxides KO_2, RbO_2, and CsO_2 form.

BTW

Metals in their elemental state have no charge. Designations such as Al^{3+} refer to the metal in, or from, a compound. If you have a piece of metal, you should use Al, or possibly Al^0.

Band Theory of Conductivity

Metals conduct electricity through conduction bands. Conduction bands arise from the application of molecular orbital theory to multi-atom systems. (See Chapter 10.) The bonding molecular orbitals and, sometimes, additional molecular orbitals merge to produce a valence band. In metals, the valence band is also the conduction band. Any partially filled band is a conduction band, providing a pathway for electron flow. This band is full for the nonmetals. A filled band cannot conduct electricity. The metalloids have a filled valence band and, at a slightly higher energy, an empty conduction band. The separation between these two bands is the band gap. A relatively small input of energy can move electrons from the filled valence band to the empty conduction band. When electrons enter the conduction band, it becomes partially filled, and electrical conductivity is possible.

Periodic Trends in Metallic Properties

Metals follow the general trends of atomic radii, ionization energy, and electron affinity. Radii increase to the left in any row and down any column on the periodic table. Ionization energies and electron affinities increase up any column and toward the right in any row on the periodic table. Electron affinities are not very important for the metals because they normally form cations. Variations appear whenever the metal has a half-filled or filled subshell of electrons. The electronegativity values for the representative metals increase toward the top of any column and toward the right on the periodic table. For the transition metals, the electronegativity peaks at gold.

The metallic properties increase down any column and toward the left in any row on the periodic table. One important metallic property is that metal oxides are base anhydrides. A base anhydride will produce a base in water. These are not oxidation-reduction reactions. Many metal oxides are too insoluble for them to produce any significant amount of base. However, most metal oxides, even those that are not soluble in water, will behave as bases to acids. A few metal oxides, and their hydroxides, are amphoteric. Amphoteric means they may behave either as a base or as an acid. Amphoterism is important for aluminum, beryllium, and zinc. Complications occur whenever the oxidation number of the metal exceeds +4 as the oxides tend to be acidic.

General Properties of Nonmetals

In a compound containing a metal and a nonmetal, the nonmetal is an anion. The anionic charge of a nonmetal is predictable from the position of the nonmetal on the periodic table. You begin on the far right and count toward the left until you get to the column containing the nonmetal of interest. The noble gases do not form anions, and this method leaves them with a zero-oxidation number. The next column to the left, the halogens, is –1; then comes –2, followed by –3, and finally –4. (Remember these are oxidation numbers; you should not expect to obtain isolated ions such as C^{4-}.) In compounds, fluorine has only a –1-oxidation number. The anionic charge determined by this procedure is the lowest possible oxidation number for the nonmetal.

All nonmetals except fluorine and the lighter noble gases (He, Ne, and Ar) can form compounds where the nonmetal has a positive oxidation number. A nonmetal can only adopt a positive oxidation number if there is a more electronegative nonmetal present. The maximum oxidation number of a nonmetal is related to the position of the nonmetal on the periodic table. Using the older system (A and B columns), the maximum oxidation is equal to the Arabic numeral for the A columns. Using the newer system, the maximum is the group number –10. Xenon exhibits the highest observed oxidation number for a nonmetal. Xenon is +8 in XeO_4.

> **EASY MISTAKE**
>
> Even though fluorine only exists in compounds in the –1 oxidation state, it, like all elements, has a zero-oxidation state in the elemental form. This means that fluorine is –1 in CaF_2, PbF_2, SF_4, and ClF_3, and 0 in F_2.

A nonmetal may adopt any oxidation number between the values predicted in the preceding two paragraphs. The only exceptions are fluorine, which is only –1 in compounds, and helium, neon, and argon, which have no known stable compounds. When there is a choice of oxidation states, there must be additional information available to allow you to choose the correct state.

Periodic Trends of Nonmetals

Nonmetals follow the general trends of atomic radii, ionization energy, and electron affinity. Radii increase to the left in any row and down any column on the periodic table. Ionization energies and electron affinities increase up any column and toward the right in any row on the periodic table. The noble gases do not have electron affinity values. Ionization energies are not very important for the nonmetals because they normally form anions. Variations appear whenever the nonmetal has a half-filled or filled subshell of electrons. The electronegativity values for the nonmetals increase toward the top of any column and toward the right on the periodic table. Fluorine has the highest electronegativity. The noble gases do not have an electronegativity value.

Chemically, nonmetals are usually the opposite of metals. The nonmetallic nature will increase toward the top of any column and toward the right in any row on the periodic table. Most nonmetal oxides are acid anhydrides. When added to water, they will form acids. A few nonmetal oxides, most notably CO and NO, do not react with water. Nonmetal oxides that do not react are neutral oxides. The reaction of a nonmetal oxide with water is not an oxidation-reduction reaction. The acid that forms will have the nonmetal in the same oxidation state as in the reacting oxide. The main exception to this is NO_2, which undergoes an oxidation-reduction (disproportionation) reaction to produce a mixture of HNO_3 and NO. When a nonmetal can form more than one oxide, the higher the oxidation number of the nonmetal, the stronger the acid it forms.

Properties of the Transition Metals

The most confusing property of the transition metals is that nearly every transition metal can adopt more than one oxidation state. In most cases, the range of oxidation states is from +2 to the group number if you use the older system (B columns) of assigning group numbers with numerals. The maximum oxidation number in the first transition series is +7 (manganese), while the highest in the second and third transition series is +8 (ruthenium and osmium). For all three series, you should expect the upper limit to increase to the maximum for the series and then start decreasing. The only +1 oxidation states are in column 1B (11) where the three metals, copper, silver, and gold, may adopt this state. The +1 state is the most stable state for silver. Zinc and cadmium use only the +2-oxidation state. The mercury(I) ion, Hg_2^{2+}, is a special case because it is a polyatomic ion. Since there are multiple oxidation states available, you must have additional information in the problem or chapter to know which one you should choose. Compounds or ions with the metal in an oxidation state above +5 are strong oxidizing agents. The ions Ti^{3+}, V^{2+}, Cr^{2+}, and Fe^{2+} are reducing agents.

Coordination Compounds: Crystal Field Theory

In addition to the ability of transition metals to adopt a variety of oxidation states, they can form coordination compounds. Coordination compounds contain complex ions. The ability to form a complex ion is not restricted to transition metals; however, most examples you will see involve a transition metal.

A complex ion will have a central atom, normally a transition metal (M), with one or more ligands (L). A ligand is a Lewis base that reacts with the central atom. The most common numbers of ligands are four and six. If a species does not contain a lone pair of electrons, it cannot be a Lewis base, which means it cannot be a ligand. Some ligands are chelating ligands. A chelating ligand behaves as a Lewis base more than once to the same central atom. An example of a chelating ligand is ethylenediamine, $NH_2CH_2CH_2NH_2$. This molecule donates a pair of electrons from each of the nitrogen atoms. Three ethylenediamine ligands count the same as six "normal" ligands. Another common chelating ligand that behaves like ethylenediamine is the oxalate ion, $C_2O_4^{2-}$. The ligand ethylenediaminetetraacetate, $EDTA^{4-}$, chelates by donating six electron pairs. The six electron pairs make this ligand equivalent to six "normal" ligands.

Crystal field theory allows predictions concerning the behavior of complexes. When six ligands are present, the complex formed is usually octahedral. If there are four ligands present, the complex formed may be either tetrahedral or square planar. Each of these three geometries has a characteristic splitting of the *d*-orbitals predicted by crystal field theory. A reference book or the internet will have diagrams of these characteristic patterns. The steps in using these patterns begin with determining the number of ligands. If the number is four, there must be additional information available for you to determine if the complex is tetrahedral or square planar. Once you have the appropriate pattern, you should enter the *d*-electrons from the metal into the pattern. If there is an option on how to do this, there must be additional information in the problem.

When a transition metal forms a cation, it is the s-electrons that leave first. Thus, iron is $[Ar]4s^2 3d^6$, and Fe^{2+} is $[Ar]3d^6$, and Fe^{3+} is $[Ar]3d^5$.

Complex Ions

The general equation for the formation of a complex is:

$$M + x\,L \rightleftharpoons [M\,L_x]$$

The equilibrium constant expression, K_f, is:

$$K_f = \frac{[ML_x]}{[M][L]^x}$$

To do a problem involving the formation of a complex ion requires you to know the value of x. While x is usually either 4 or 6, it can have other values. How can you tell what x is? You must have additional information. A problem may tell you directly what x will be. In other cases, you may need to examine a table of formation constants, K_f. If you see, for example, K_f for $[Fe(C_2O_4)_3]^{3-}$ is equal to some value, the subscript "3" tells you that $x = 3$.

To change the general equation to a specific equation entails a simple substitution of the formulas. Using $[Fe(C_2O_4)_3]^{3-}$ as an example, we see that $M = Fe^{3+}$ and $L = C_2O_4^{2-}$, to give the equation:

$$Fe^{3+}(aq) + 3\ C_2O_4^{2-}(aq) \leftrightarrows [Fe(C_2O_4)_3]^{3-}(aq)$$

This leads to the following equilibrium constant expression:

Make sure you move the ion charges inside the bracket, so they are not mistaken for exponents.

$$K_f = \frac{\left[Fe\left(C_2O_4\right)_3^{3-}\right]}{\left[Fe^{3+}\right]\left[C_2O_4^{2-}\right]^3}$$

Once you have the correct equilibrium constant expression, there is no difference in solving the complex ion equilibrium than any other equilibrium problem.

EXERCISES

EXERCISE 19-1

Answer the following questions.

1. What element is nearly always an exception to all the "rules" of chemistry?

2. What is the oxidation state of hydrogen when it combines with a metal?

3. What is the oxidation number of Os in OsO_4?

4. Which elements form superoxides?

5. Metals above hydrogen on the activity series react with acids to produce what gas?

6. Which transition metal has the highest electronegativity of all transition metals?

7. Based on its position on the periodic table, what are the highest and lowest oxidation numbers for phosphorus?

8. In the formation of a complex, which substance serves as a Lewis base?

9. Write the general equation for the formation of a complex.

10. Write the general equation for an equilibrium constant concerning the formation of a complex.

EXERCISE 19-2

Write balanced chemical equations for the reaction of each of the following with water.

1. $Na_2O(s)$

2. $CaO(s)$

3. $SO_2(g)$

4. $N_2O_5(s)$

5. $NO(g)$

Flashcard App

Nuclear Chemistry

MUST KNOW

⚡ Nuclear decay reactions involve the transformation of an unstable isotope into a more stable one.

⚡ The half-life ($t_{1/2}$) is the amount of time that it takes for one-half of a sample of a radioactive isotope to decay. This decay follows first-order kinetics.

⚡ Einstein's equation lets us calculate the amount of energy released when a certain amount of matter is converted to energy.

In this chapter we're going to look at nuclear reactions, including nuclear decay as well as fission and fusion. If you need to, review the section in Chapter 2 on isotopes and the section in Chapter 13 on integrated rate laws, which discusses first-order kinetics. Let's go.

Nuclear Reactions

Most nuclear reactions involve the breaking apart of the nucleus into two or more different elements and/or subatomic particles. If we know all but one of the particles, then the unknown particle can be determined by balancing the nuclear equation. When chemical equations are balanced, we add coefficients to ensure there are the same number of each type of atom on both the left and right of the reaction arrow. However, to balance nuclear equations, we ensure there is the same sum of both mass numbers and atomic numbers on the left and right of the reaction arrow. Recall that we can represent a specific isotope of an element by the following symbolization:

$$^A_Z X$$

A is the mass number (sum of protons and neutrons), Z is the atomic number (number of protons), and X is the element symbol (from the periodic table). In balancing nuclear reactions, ensure that the sum of all A values on the left of the reaction arrow equals the sum of all A values to the right of the arrow. The same will be true of the sums of the atomic numbers, Z. Knowing that these sums must be equal allows you to predict the mass and atomic number of an unknown particle, if we know all the others.

EXAMPLE

▶ If we bombard chlorine-35 with a neutron, we create hydrogen-1 along with an isotope of a different element. Write a balanced nuclear reaction for this process.

▶ First, we write a partial nuclear equation:

$$^{35}_{17}\text{Cl} + ^{1}_{0}n \longrightarrow ^{1}_{1}\text{H} + ^{x}_{y}?$$

▶ The sum of the mass numbers on the left of the equation is 36 = (35 + 1) and on the right is (1 + x). The mass number of the unknown isotope must be 35. The sum of the atomic numbers on the left is 17 = (17 + 0) and (1 + y) on the right. The atomic number of the unknown must then be 16. This atomic number identifies the element as sulfur, so we can write a complete nuclear equation:

BTW

Sulfur-35 does not occur in nature; it is an artificial isotope.

$$^{35}_{17}\text{Cl} + ^{1}_{0}n \longrightarrow ^{1}_{1}\text{H} + ^{35}_{16}\text{S}$$

We can observe three common types of radioactive decay in nature. We can occasionally observe others.

Alpha Emission

An alpha particle is essentially a helium nucleus with two protons and two neutrons. It is represented as $^{4}_{2}\text{He}$ or $^{4}_{2}\alpha$. As this particle leaves the decaying nucleus, it has no electrons and thus has a 2+ charge. However, it quickly acquires two electrons from the surroundings to form the neutral atom. Most commonly, we show the alpha particle as the neutral particle and not the cation.

Radon-222 undergoes alpha decay according to the following balanced equation:

$$^{222}_{86}\text{Rn} \longrightarrow ^{218}_{84}\text{Po} + ^{4}_{2}\text{He}$$

Notice that in going from Rn-222 to Po-218, the atomic number has decreased by 2 and the mass number by 4.

Beta Emission

A beta particle is essentially an electron and can be represented as either $_{-1}^{0}\beta$ or $_{-1}^{0}e$. This electron comes from the nucleus and not the electron cloud. It results from the conversion of a neutron into a proton and an electron:

$$_{0}^{1}n \rightarrow {}_{1}^{1}p + {}_{-1}^{0}e$$

Nickel-63 will undergo beta decay according to the following equation:

$$_{28}^{63}\text{Ni} \rightarrow {}_{29}^{63}\text{Cu} + {}_{-1}^{0}e$$

Notice that the atomic number has increased by one in going from nickel-63 to copper-63 but the mass number has remained unchanged.

Gamma Emission

Gamma emission is the release of high-energy, short-wavelength photons, which are like x-rays. The representation of this radiation is γ. Gamma emission commonly accompanies most other types of radioactive decay, but we normally do not show it in the balanced nuclear equation, since it has neither appreciable mass nor charge.

Alpha, beta, and gamma emission are the most common types of natural decay modes, but we do occasionally observe positron emission and electron capture.

Positron Emission

A positron is essentially an electron that has a positive charge instead of a negative one. It is represented as $_{+1}^{0}\beta$ or $_{+1}^{0}e$. Positron emission results from the conversion of a proton to a neutron and a positron:

$$_{1}^{1}p \rightarrow {}_{0}^{1}n + {}_{+1}^{0}e$$

We observe it in the decay of some radioactive isotopes, such as potassium-40:

$$^{40}_{19}K \rightarrow {}^{40}_{18}Ar + {}^{0}_{+1}e$$

Electron Capture

The four decay modes described previously all involve emission or giving off a particle, but electron capture is the capturing of an electron from the energy level closest to the nucleus (1s) by a proton in the nucleus. This creates a neutron:

$$^{1}_{1}p + {}^{0}_{-1}e \rightarrow {}^{1}_{0}n$$

However, this leaves a vacancy in the 1s energy level and an electron, from a higher energy level, drops down to fill this vacancy. A cascading effect occurs as the electrons shift downward, releasing energy. This released energy falls in the x-ray part of the electromagnetic spectrum. These x-rays give scientists a clue that electron capture has taken place.

Polonium-204 undergoes electron capture:

$$^{204}_{84}Po + {}^{0}_{-1}e \rightarrow {}^{204}_{83}Bi + x\text{-rays}$$

Notice that the atomic number has decreased by 1, but the mass number has remained the same. This is the same result as for positron emission.

BTW

Electron capture is the only decay process we presented where you add a particle on the left side of the reaction arrow.

Nuclear Stability

Sometimes it is difficult to predict if an isotope is stable and, if unstable, what type of decay mode it might undergo. All isotopes that contain 84 or more protons are unstable. These unstable isotopes will undergo nuclear decay. For these massive isotopes, we observe alpha decay most commonly. Alpha decay gets rid of four units of mass and two units of charge, thus helping to relieve the repulsive stress found in the nucleus of these isotopes. For other isotopes of atomic number less than 83, we can best predict stability using the neutron to proton (n/p) ratio.

A plot of the number of neutrons (n) versus the number of protons ($p = Z$) for the known stable isotopes gives the nuclear belt of stability. (See the internet for a figure of the belt or region of stability.) At the low end of this belt of stability ($Z < 20$), the n/p ratio is 1. At the high end ($Z < 80$), the n/p ratio is about 1.5. We can then use the n/p ratio of the isotope to predict if it will be stable. If it is unstable, then the isotope will use a decay mode that will bring it back onto the belt of stability.

Consider neon-18 or Ne-18. It has 10 p and 8 n, giving an n/p ratio of 0.8. For a light isotope, like this one, this value is low. A low value indicates that this isotope will probably be unstable. Neutron-poor isotopes, meaning that they have a low n/p ratio, do not have enough neutrons (or have too many protons) to be stable. Decay modes that increase the number of neutrons and/or decrease the number of protons are favorable. Both positron emission and electron capture accomplish this by converting a proton into a neutron. In general, positron emission occurs with lighter isotopes and electron capture with heavier ones.

Isotopes that are neutron-rich, that have too many neutrons or not enough protons, lie above the belt of stability and tend to undergo beta emission because that decay mode converts a neutron into a proton.

A particular isotope may undergo a series of nuclear decays until finally a stable isotope forms. For example, radioactive U-238 decays to stable Pb-206 in 14 steps; half of these are alpha emissions and the other half are beta emissions.

While the position of an isotope on the neutron-proton plot indicates what is the more likely decay mode, it is not a guarantee that that mode will occur.

Half-Lives ($t_{1/2}$)

A radioactive isotope is unstable, but it is impossible to predict when a certain atom will decay. However, if we have a statistically large enough sample, some trends become obvious. The radioactive decay follows first-order kinetics (see Chapter 13 for a more in-depth discussion of first-order reactions). If we monitor the number of radioactive atoms in a sample, we observe that it takes a certain amount of time for half the sample to decay; it takes the same amount of time for half the remaining sample to decay, and so on. The amount of time it takes for half the sample to decay is the half-life of the isotope and has the symbol $t_{1/2}$. The following table shows the percentage of the radioactive isotope remaining versus half-life.

Half-Life, $t_{1/2}$	Percent Radioactive Isotope Remaining
0	100.000
1	50.000
2	25.000
3	12.500
4	6.250
5	3.125
6	1.562
7	0.781
8	0.391
9	0.195
10	0.098

Half-lives may be very short, 4.2×10^{-6} seconds for Po-213, or very long, 4.5×10^9 years for U-238.

The half-life of an isotope is a consideration when that isotope is used for medical purposes. Iodine-131 is used to treat an overactive thyroid gland since it is a beta emitter and concentrates in the thyroid. Thankfully, it has a half-life of only eight days. It can accomplish its task, but not be active in the body for a long time.

If only multiples of half-lives are considered, the calculations are very straightforward.

EXAMPLE

▶ Iodine-131 is useful in the treatment of thyroid cancer and has a $t_{1/2}$ of 8 days. How long would it take to decay to 25% of its original amount?

▶ Looking at the preceding chart, we can see that 25% would be at 2 half-lives or 16 days.

Since radioactive decay is not a linear process, you cannot use the chart to predict how much would still be radioactive at the end of some time (or amount) that is not associated with a multiple of a half-life. To solve these types of problems, one must use the mathematical relationships associated with first-order kinetics. In general, two equations from Chapter 13 are useful:

$$\ln (N_t/N_0) = -kt \qquad\qquad t_{1/2} = (\ln 2)/k \text{ or } t_{1/2} = 0.693/k$$

In these equations, ln is the natural logarithm, N_t is the amount of radioactive isotope at some time t, N_0 was the amount of radioactive isotope initially present, and k is the rate constant for the decay. If you know initial and final amounts and if you are looking for the half-life, you would use the first equation to solve for the rate constant and then use the second to solve for $t_{1/2}$ as done in the following example.

EXAMPLE

▶ What is the half-life of a radioisotope that takes 15 min to decay to 90% of its original activity?

▶ Using our first equation:

$$\ln (90/100) = -k\,(15\ \text{min})$$

$$-0.1054 = -k\,(15\ \text{min})$$

$$7.03 \times 10^{-3}\ \text{min}^{-1} = k$$

▶ Now the second equation:

$$t_{1/2} = \ln 2/7.03 \times 10^{-3}\text{min}^{-1}$$

$$t_{1/2} = 0.693/7.03 \times 10^{-3}\ \text{min}^{-1}$$

$$t_{1/2} = 98.5775 = 98.6\ \text{min}$$

If one knows the half-life and amount remaining radioactive, you can then use equation (2) to calculate the rate constant, k, and then use equation (1) to solve for the time. This is the basis of carbon-14 dating. Scientists use carbon-14 dating to determine the age of objects that were once alive.

EXAMPLE

▶ Suppose we discover a wooden tool, and we determine its carbon-14 activity to have decreased to 65% of the original. How old is the object?

▶ The half-life of C-14 is 5730 years. Substituting this into the $t_{1/2}$ equation:

$$5730\ \text{y} = (\ln 2)/k$$

$$5730\ \text{y} = 0.693/k$$

$$k = 1.21 \times 10^{-4}\ \text{y}^{-1}$$

▶ Substituting this rate constant into the rate-constant equation:

$$\ln (65/100) = -(1.21 \times 10^{-4} \, \text{y}^{-1}) \, t$$

$$-0.4308 = -(1.21 \times 10^{-4} \, \text{y}^{-1}) \, t$$

$$t = 3560.33 = 3560 \text{ years}$$

Mass/Energy Conversions

Whenever a nuclear decay or reaction takes place, it releases energy. This energy may be in the form of heat and light, gamma radiation, or kinetic energy of the expelled particle and recoil of the remaining particle. This energy results from the conversion of a very small amount of matter into energy.

When dealing with energy in nuclear processes, it is important to consider the mass defect. The **mass defect** is the difference in the mass of a nuclear and the sum of the masses of the components of that nucleus. For example, the molar mass of a helium-4 atom is 4.002603 g/mol, while the mass of two protons plus two neutrons plus two electrons is 4.032981 g/mol. The difference between these two values (0.030378 g/mol) is the mass defect. The mass of all nuclei except hydrogen-1 is less than the sum of the masses of the components.

BTW

In a nuclear reaction, there is no conservation of mass as in ordinary chemical reactions.

The amount of energy that is produced can be calculated by using Einstein's relationship, $E = mc^2$, where E is the energy produced, m is the mass converted into energy (the mass defect), and c is the speed of light. The amount of matter converted into energy is normally very small, but when we multiply it by the speed of light (a very large number) squared, the amount of energy produced is very large.

When 1 mole of U-238 decays to Th-234, 5×10^{-6} kg of matter is converted to energy (the mass defect). Calculate the amount of energy released.

BTW

If you want to calculate the energy in joules, the mass must be in units of kilograms and the speed of light must have units of m/s. A joule is a kg·m²/s².

$$E = mc^2$$

$$E = (5 \times 10^{-6}\ \text{kg})(3.00 \times 10^8\ \text{m/s})^2$$

$$E = 5 \times 10^{11}\ \text{kg} \cdot \text{m}^2/\text{s}^2 = 4.5 \times 10^{11} = 5 \times 10^{11}\ \text{J}$$

Fission and Fusion

Nuclear fission is the breakdown of a nucleus randomly into two or more smaller nuclei with the release of energy. The most useful fission process involves the decay of U-235 when hit by a neutron, and one example of this transformation is:

$$^{1}_{0}n + ^{235}_{92}U \longrightarrow ^{142}_{56}Ba + ^{91}_{36}Kr + 3\ ^{1}_{0}n$$

Notice that the reaction consumes one neutron, but the reaction releases three neutrons. Those three neutrons are then free to initiate additional fission reactions. This type of situation in which there is a multiplier effect is a chain reaction. We can use isotopes that undergo chain reaction in both the production of bombs and in nuclear power plants. U-235 is fissionable, but U-238 is not. There is a certain minimum quantity of fissionable matter needed to support a chain reaction, the critical mass.

Nuclear reactors are useful in the production of electricity, but they are not without their problems. These problems include disposal of nuclear wastes, accidents, and sabotage. The eventual answer may lie in nuclear fusion.

Fusion is the combining of lighter nuclei into a heavier one. Such reactions can release a great deal of energy. Isotopes of hydrogen fuse into helium and power the sun. For the past few decades, scientists have been investigating the fusion process as a way of providing the world with energy. One of the more promising fusion reactions is:

$$\mathstrut_{1}^{2}H + \mathstrut_{1}^{3}H \rightarrow \mathstrut_{2}^{4}He + \mathstrut_{0}^{1}n$$

In this reaction, two isotopes of hydrogen fuse into helium and a neutron. Four major problems in this reaction have arisen—time, temperature, containment, and the neutrons produced. The nuclei must be held together long enough (~1 s) at high enough temperatures to provide the activation energy for the reaction (~40,000,000 K). At this temperature, every substance is a gas or plasma, so containment is proving to be the biggest obstacle. Scientists are conducting major investigations into the use of magnetic fields ("magnetic bottles") to contain the nuclei at this temperature. The neutrons formed will not be contained by the magnetic bottle, because they have no charge. These neutrons are free to be absorbed by other nuclei in the area to produce radioactive isotopes. If they can overcome these obstacles, fusion may well provide a limitless energy source for our world.

Nuclear Decay

To solve nuclear decay problems, we use one or both equations we've just been working with:

$$\ln (N_t/N_0) = -kt \qquad\qquad t_{1/2} = (\ln 2)/k \text{ or } t_{1/2} = 0.693/k$$

The terms N_t and N_0 may have many different units. The only restriction on the units is that these two terms have the same units. If they do not have the same units, you will need to convert one of the units. Both t and $t_{1/2}$ will have units of time. Any time unit may be present. The rate constant (decay constant), k, will have units of time^{-1}. Any time unit may be present. The time units for t, $t_{1/2}$, and k must agree. If the units do not agree, you will need to make a conversion.

Let's look at some sample questions.

EXAMPLE

▶ Determine the decay constant for iron-55. The half-life of iron-55 is 2.7 years.

▶ You should begin a problem of this type by recopying and labeling each of the known and unknown variables:

$$k = ?$$

$$t_{1/2} = 2.7 \text{ years}$$

▶ These two variables appear in the equation $t_{1/2} = (\ln 2)/k$. We can rearrange this equation to find k, and enter the value for the half-life:

$$k = (\ln 2)/t_{1/2} = (\ln 2)/2.7 \text{ y} = 0.693/2.7 \text{ y} = 0.256667 = 0.26 \text{ y}^{-1}$$

Let's try another twist by using a decay constant to calculate a half-life.

EXAMPLE

▶ What is the half-life for argon-41? The decay constant for argon-41 is $6.33 \times 10^{-3} \text{ min}^{-1}$.

▶ We can recopy and label the variables to get:

$$k = 6.33 \times 10^{-3} \text{ min}^{-1}$$

$$t_{1/2} = ?$$

▶ These two variables appear in the equation $t_{1/2} = (\ln 2)/k$. We do not need to rearrange this equation. We only need to enter the value for the decay constant.

$$t_{1/2} = (\ln 2)/k = (\ln 2)/6.33 \times 10^{-3} \text{ min}^{-1}$$
$$= 0.693 / 6.33 \times 10^{-3} \text{ min}^{-1}$$

$$= 109.47867 = 109 \text{ min}$$

We have been concentrating on how long, so let's work on mass now.

EXAMPLE

▶ A sample of iron-55 weighs 3.75 mg. The decay constant for this isotope is 0.26 y^{-1}. How much iron-55 will remain in the sample after 6.0 years?

▶ We can recopy and label the variables to get:

$N_0 = 3.75$ mg

$N_t = ?$

$k = 0.26$ y^{-1}

$t = 6.0$ y

▶ The only equation containing these four variables is $\ln (N_t/N_0) = -kt$. We can enter the values into the equation to get:

$$\ln \left(\frac{?}{3.75 \text{ mg}} \right) = - (0.26 \text{ y}^{-1}) (6.0 \text{ y}) = - 1.56 \text{ (unrounded)}$$

$$\left(\frac{?}{3.75 \text{ mg}} \right) = e^{-1.56} = 0.210136 \text{ (unrounded)}$$

$$? = (0.210136) (3.75 \text{ mg}) = 0.78801 = 0.79 \text{ mg}$$

In the previous problem we knew how much we started with and found how much would be left at a certain time. Now let's see if we can reverse that and calculate how much we start with.

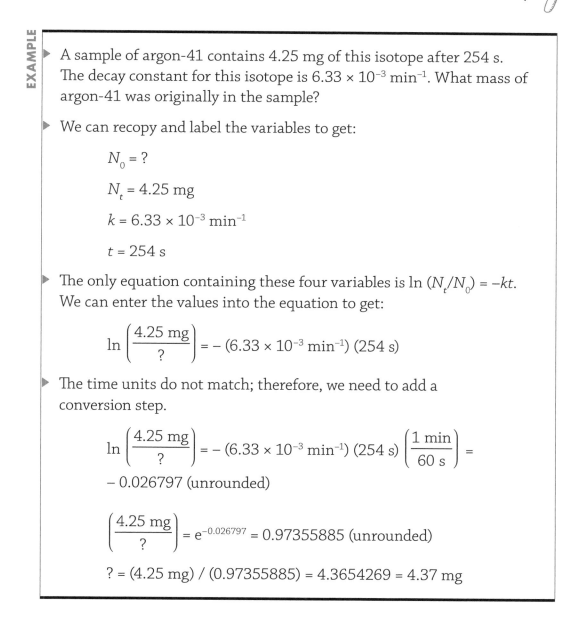

EXAMPLE

▶ A sample of argon-41 contains 4.25 mg of this isotope after 254 s. The decay constant for this isotope is 6.33×10^{-3} min^{-1}. What mass of argon-41 was originally in the sample?

▶ We can recopy and label the variables to get:

$$N_0 = ?$$

$$N_t = 4.25 \text{ mg}$$

$$k = 6.33 \times 10^{-3} \text{ min}^{-1}$$

$$t = 254 \text{ s}$$

▶ The only equation containing these four variables is $\ln (N_t/N_0) = -kt$. We can enter the values into the equation to get:

$$\ln \left(\frac{4.25 \text{ mg}}{?} \right) = - (6.33 \times 10^{-3} \text{ min}^{-1}) (254 \text{ s})$$

▶ The time units do not match; therefore, we need to add a conversion step.

$$\ln \left(\frac{4.25 \text{ mg}}{?} \right) = - (6.33 \times 10^{-3} \text{ min}^{-1}) (254 \text{ s}) \left(\frac{1 \text{ min}}{60 \text{ s}} \right) =$$

$$- 0.026797 \text{ (unrounded)}$$

$$\left(\frac{4.25 \text{ mg}}{?} \right) = e^{-0.026797} = 0.97355885 \text{ (unrounded)}$$

$$? = (4.25 \text{ mg}) / (0.97355885) = 4.3654269 = 4.37 \text{ mg}$$

One of the most valuable uses of radioactive decay is in radiocarbon dating. The following example will show how this works.

▶ The current decay rate of the ^{14}C in a sample is 4.82 disintegrations per minute per gram (d/min • g). The ^{14}C activity of living organisms is 15.3 d/min • g. The half-life of ^{14}C is 5730 years. How many years old is the sample?

▶ We can recopy and label the variables to get:

$$N_0 = 15.3 \text{ d/min} \cdot \text{g}$$

$$N_t = 4.82 \text{ d/min} \cdot \text{g}$$

$$t_{1/2} = 5730 \text{ y}$$

$$t = ?$$

▶ There is no single equation containing these four variables. For this reason, we need to use the two equations: $\ln(N_t/N_0) = -kt$ and $t_{1/2} = (\ln 2)/k$. We will begin with the half-life equation $t_{1/2} = (\ln 2)/k$. We need to rearrange this equation and enter the half-life to determine the decay constant:

$$k = (\ln 2)/t_{1/2} = (\ln 2)/5730 \text{ y} = 0.693/5730 \text{ y} = 1.209424 \times 10^{-4} \text{ y}^{-1} \text{ (unrounded)}$$

▶ Now that we have the value of the decay constant, we can use the other equation. We rearrange this equation:

$$t = -\frac{\ln\left(\dfrac{N_t}{N_0}\right)}{k}$$

▶ We can now enter the values and calculate the time:

$$t = -\frac{\ln\left(\dfrac{4.82 \text{ d}/\min \cdot \text{g}}{15.3 \text{ d}/\min \cdot \text{g}}\right)}{1.209402 \times 10^{-4} \text{ y}^{-1}} = 9550.8 = 9.55 \times 10^3 \text{ y}$$

EXERCISES

EXERCISE 20-1

Answer these questions about nuclear chemistry.

1. A typical periodic table always lists which of the following? (There may be more than one correct answer.)
    a. chemical symbol
    b. atomic number
    c. mass number
    d. atomic weight
    e. physical state

2. Two atoms with the same number of protons but different numbers of neutrons are examples of _____ of an element.

3. Write symbols for each of the following.
    a. alpha particle
    b. beta particle
    c. gamma ray
    d. positron

4. Balance the following nuclear equations. Gamma rays are not necessary.
    a. $^{137}_{58}\text{Ce} + \underline{\hspace{1.5cm}} \rightarrow \, ^{137}_{57}\text{La}$

    b. $^{205}_{80}\text{Hg} \rightarrow \, ^{205}_{81}\text{Tl} + \underline{\hspace{1.5cm}}$

    c. $^{245}_{96}\text{Cm} \rightarrow \, ^{241}_{94}\text{Pu} + \underline{\hspace{1.5cm}}$

    d. $^{256}_{103}\text{Lr} \rightarrow \underline{\hspace{1.5cm}} + \, ^{4}_{2}\text{He}$

    e. $^{238}_{92}\text{U} + \, ^{1}_{0}n \rightarrow \underline{\hspace{1.5cm}}$

5. If 6.25% of a radioactive isotope remains after 40 days, what is the half-life of the isotope?

6. Define:

   a. fission

   b. fusion

7. The mass defect for helium-4 is 0.0304 g/mol. Determine the nuclear binding energy in joules per mole for 1 mole of helium-4.

8. The half-life of krypton-85 is 10.76 years. What percentage of a sample of krypton-85 will remain after 21.52 years?

9. After 182 min, the amount of arsenic-78 in a sample has decreased from 5.00 mg to 1.25 mg. What is the half-life of arsenic-78?

10. A sample of hair from an Egyptian mummy gives off radiation from carbon-14 at a rate of 58.2% of a present-day sample. How old is the mummy if the half-life of carbon-14 is 5730 years?

Flashcard App

Organic Chemistry, Biochemistry, and Polymers

MUST KNOW

- Organic chemistry is the study of carbon-containing compounds. Hydrocarbons contain only carbon and hydrogen.

- Alkanes contain only carbon-to-carbon single bonds, while alkenes contain one or more carbon-to-carbon double bonds and alkynes contain one or more triple bonds.

- Functional groups are groups in organic compounds that give the molecule its characteristic reactivity.

- Polymers are large, high molecular weight compounds formed by linking together smaller monomer units. These include proteins and carbohydrates.

- Proteins are naturally occurring polymers of amino acid monomer units joined by a peptide bond. Carbohydrates are composed of carbon, hydrogen, and oxygen. Cellulose and starch are examples of carbohydrates.

Organic compounds, most compounds containing carbon, are extremely important in our daily lives. They are also extremely abundant. Almost 90 percent of the 60 million or so known chemical compounds are organic compounds. We are composed of organic compounds; we consume organic compounds; we wear organic compounds. When we talk about biochemistry, we are talking about organic compounds, and polymers (plastics), whether natural or synthetic, are made of organic compounds. In this chapter we will delve into the world of carbon compounds.

Organic Compounds

Organic chemistry is the study of the chemistry of carbon. We classify almost all the compounds containing carbon as organic compounds. We consider only a few, such as carbonates, cyanides, and so on, as inorganic. Scientists once believed that only living organisms could produce organic compounds. However, in 1828, the German chemist Friedrich Wöhler showed this to be incorrect when he produced the first organic compound from inorganic starting materials. From that time, chemists have synthesized many organic compounds found in nature and have made many never found naturally. Carbon is capable of strongly bonding to itself and other elements. There appears to be no limit to how many carbon atoms can bond together. These factors allow carbon to form long, complex chains and rings.

Hydrocarbons and Nomenclature

Alkanes, alkenes, alkynes, and aromatic compounds are members of a family of organic compounds called hydrocarbons, compounds of carbon and hydrogen.

More mistakes in the study of organic chemistry can be related to forgetting one simple fact: carbon forms four bonds.

IRL Hydrocarbons are the simplest of organic compounds but are extremely important to our society as fuels and raw materials for chemical industries. We heat our homes and run our automobiles through the combustion (burning) of these hydrocarbons, and plastic, synthetic fabrics, most medicines, and so on are all made from hydrocarbons.

Alkanes are hydrocarbons that contain only single covalent bonds within the molecule. They are saturated hydrocarbons since the carbon atoms bond to the maximum number of other atoms. These alkanes may be straight-chained hydrocarbons, in which the carbons are sequentially bonded, branched hydrocarbons in which another hydrocarbon group is bonded to the hydrocarbon "backbone," or they may be cyclic in which the hydrocarbon is composed entirely or partially of a ring system. The straight-chained and branched alkanes have the general formula of C_nH_{2n+2}, while the cyclic alkanes will have the general formula of C_nH_{2n}. The n stands for the number of carbon atoms in the compound. The first 10 straight-chained hydrocarbons are shown in the following table:

The First 10 Straight-Chained Hydrocarbons

Name	Molecular Formula	Structural Formula
Methane	CH_4	CH_4
Ethane	C_2H_6	$CH_3\text{-}CH_3$
Propane	C_3H_8	$CH_3\text{-}CH_2\text{-}CH_3$
Butane	C_4H_{10}	$CH_3\text{-}CH_2\text{-}CH_2\text{-}CH_3$
Pentane	C_5H_{12}	$CH_3\text{-}CH_2\text{-}CH_2\text{-}CH_2\text{-}CH_3$
Hexane	C_6H_{14}	$CH_3\text{-}CH_2\text{-}CH_2\text{-}CH_2\text{-}CH_2\text{-}CH_3$
Heptane	C_7H_{16}	$CH_3\text{-}CH_2\text{-}CH_2\text{-}CH_2\text{-}CH_2\text{-}CH_2\text{-}CH_3$
Octane	C_8H_{18}	$CH_3\text{-}CH_2\text{-}CH_2\text{-}CH_2\text{-}CH_2\text{-}CH_2\text{-}CH_2\text{-}CH_3$
Nonane	C_9H_{20}	$CH_3\text{-}CH_2\text{-}CH_2\text{-}CH_2\text{-}CH_2\text{-}CH_2\text{-}CH_2\text{-}CH_2\text{-}CH_3$
Decane	$C_{10}H_{22}$	$CH_3\text{-}CH_2\text{-}CH_2\text{-}CH_2\text{-}CH_2\text{-}CH_2\text{-}CH_2\text{-}CH_2\text{-}CH_2\text{-}CH_3$

There can be many more carbon units in a chain than are shown in the preceding table, but those are enough to allow us to study a little alkane nomenclature, the naming of alkanes.

The naming of alkanes relies on choosing the longest carbon chain in the structural formula. Then name the hydrocarbon branches with a number indicating to which carbon atom the branch is attached. Here are the specific rules for naming simple alkanes:

Find the continuous carbon chain in the compound that contains the most carbon atoms.	This will be the base name of the alkane.
You will modify this base name by adding the names of the branches (substituent groups) in front of the base name.	We name alkane branches by taking the name of the alkane that contains the same number of carbon atoms, dropping the *-ane* ending, and adding *-yl*. Methane would then become methyl; propane becomes propyl, and so on. If there is more than one branch, list them alphabetically.
A location number is necessary to indicate the point of attachment of a substituent.	We assign these numbers by consecutively numbering the carbons of the base hydrocarbon starting at one end of the hydrocarbon chain. Choose the end so that there will be the lowest sum of location numbers for the substituent groups. Place this location number in front of the substituent name and separate it from the name by a hyphen (i.e., 2-methyl).
Place the substituent names with their location numbers in front of the base name of the alkane in alphabetical order.	If there are identical substituents (e.g., two methyl groups), then give the location numbers of each, separated by commas using the common Greek prefixes (such as, *di-*, *tri-*, and *tetra-*) to indicate the number of identical substituent groups (i.e., 2,3-dimethyl). Do not use these Greek prefixes in the alphabetical arrangement.
The last substituent group becomes a part of the base name as a prefix.	

Compounds that have the same molecular formulas, but different structural formulas, are isomers. When dealing with hydrocarbons, this amounts to a different arrangement of the carbon atoms. Isomers such as these are structural isomers.

EASY MISTAKE

In writing structural isomers as well as any organic compound, remember that carbon forms four bonds! One of the most common mistakes that a chemistry student makes is writing an organic structure with a carbon atom having less or more than four bonds!

EXAMPLE

▶ Name the following compound:

$$
\begin{array}{c}
\qquad\quad CH_3 \\
\qquad\quad | \\
CH_3-C\!\!-\!\!CH_2-CH_2-CH-CH_2-CH_3 \\
\qquad\quad | \qquad\qquad\quad | \\
\qquad\quad CH_3 \qquad\qquad CH_2 \\
\qquad\qquad\qquad\qquad\quad | \\
\qquad\qquad\qquad\quad CH_2-CH_2-CH_3
\end{array}
$$

▶ First, pick the longest chain. This is shown in the following diagram in boldface. Since only single bonds are present between the carbon atoms, this is an alkane. Since the longest chain has nine carbons, this is a nonane.

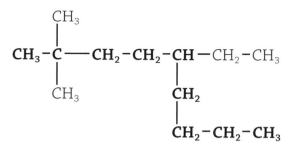

BTW

In the structure to the side, we could have chosen either of the other two -CH₃ groups to be the first carbon on the left-hand side.

▸ Next, number the longest chain from one end to the other with the lowest number(s) going to the branches. For the preceding example the numbering of the chain (boldface carbon atoms) would be:

```
1 2 3 4 5
        6
        7 8 9
```

▸ Once you have assigned these numbers, do not alter them later.

▸ All carbon atoms that are not part of the nine-atom main chain are branches. Branches have -*yl* endings. It may help you to circle the carbon atoms belonging in the branches. In the preceding example, there are three branches. Two consist of only one carbon and are methyl groups. The remaining branch has two carbon atoms, so it is an ethyl group. We arrange the branches alphabetically. If there is more than one of a type, use a prefix (such as *di-*, *tri-*, and *tetra-*). The two methyl groups are designated dimethyl. We indicate the position of each branch with a number already determined from the main chain. Each branch must get its own number even if it is identical to one already used. This gives 5-ethyl-2,2-dimethylnonane:

- ethyl before methyl (alphabetical, prefixes are ignored)
- two methyl groups → dimethyl
- three branches → three numbers

▸ We separate numbers from other numbers by commas, and we separate numbers from letters by a dash.

Alkenes are hydrocarbons that have at least one carbon-to-carbon double bond, while alkynes have at least one carbon-to-carbon triple bond. Alkenes have the general formula of C_nH_{2n} while the alkynes have the general formula of C_nH_{2n-2}. Cyclic alkenes and alkynes would have two less hydrogen atoms. Aromatic hydrocarbons are usually ring systems of alternating double and single bonds. Benzene, C_6H_6, is a very common aromatic hydrocarbon. The presence of a double or triple bond makes these hydrocarbons unsaturated; that is, they do not have the maximum number of bonds to other atoms.

Benzene, C_6H_6, is the best-known aromatic compound. It consists of a ring of six carbon atoms. One way of representing the structure is to alternate single and double bonds about the ring as shown in the leftmost diagram in the following figure. The center structure is a resonance form of the first structure. The presence of resonance stabilizes the structure by delocalizing the electrons. Resonance makes all the carbon-carbon bonds equal instead of some single and some double. The structure at the far right in the following diagram is a common representation of benzene to indicate the presence of resonance along with the equality of all carbon-carbon bonds.

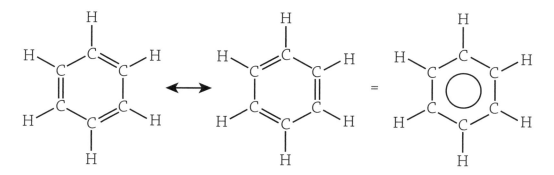

The nomenclature of alkenes and alkynes is very similar to that of the alkanes with two major differences:

- The longest carbon chain must contain the double or triple bond and number the chain from the end closest to the double/triple bond. A number indicates the position of the double/triple bond. This number indicates the number of the first carbon atom that is part of the double/triple bond.

- The name of longest carbon chain is formed by taking the alkane name, dropping the *-ane* suffix, and replacing it by *-ene* for an alkene or *-yne* for an alkyne.

$$CH_2=CH—CH_2—CH_3 \qquad\qquad CH_3—CH_2—C\equiv C—CH_3$$

1-butene   2-pentyne

Functional Groups

If chemistry students had to learn the properties of each of the millions of organic compounds, it would be an impossible task. However, chemists find that having certain arrangements of atoms in an organic molecule causes those molecules to react in a similar fashion. For example, methyl alcohol, $CH_3—OH$, and ethyl alcohol, $CH_3—CH_2—OH$, undergo the same types of reaction. The $—OH$ group is the reactive part of these types of molecules. These reactive groups are functional groups. Instead of learning the properties of individual molecules, we can simply learn the properties of functional groups.

In our study of the simple hydrocarbons, there are only two functional groups. One is a carbon-to-carbon double bond. Hydrocarbons that contain a carbon-to-carbon double bond are alkenes. The other hydrocarbon functional group is a carbon-to-carbon triple bond. Hydrocarbons that contain a triple bond are alkynes. These functional groups are the reactive sites in the alkenes and alkynes. The result is that alkenes and alkynes are more reactive than the alkanes.

The introduction of other atoms (such as N, O, and Cl) to the organic compounds gives rise to many other functional groups. The most common functional groups are in the table Common Organic Functional Groups in the next section. These functional groups normally are the reactive sites in the molecules.

Polymers

Polymers are large, high-molecular weight compounds formed by linking together many smaller compounds called monomers. The properties of the polymer are dependent on the monomer units used and the way in which they link together. Many polymers occur in nature such as cellulose, starch, cotton, wool, and rubber. Others are created synthetically, such as nylon, PVC, polystyrene, Teflon, and polyester.

Common Organic Functional Groups

Functional Group	Compound Type	Suffix or Prefix of Name	Example	Systematic Name (Common Name)
$>C=C<$	alkene	-ene	(structure) ethene	ethene (ethylene)
$-C\equiv C-$	alkyne	-yne	$H-C\equiv C-H$	ethyne (acetylene)
$-\overset{\mid}{\underset{\mid}{C}}-\ddot{\underset{\cdot\cdot}{O}}-H$	alcohol	-ol	(structure)	methanol (methyl alcohol)
$-\overset{\mid}{\underset{\mid}{C}}-\ddot{\underset{\cdot\cdot}{X}}:$ (X=halogen)	haloalkane	halo-	(structure)	chloromethane (methyl chloride)
$-\overset{\mid}{\underset{\mid}{C}}-\overset{\mid}{\underset{\mid}{N}}-$	amine	-amine	(structure)	ethylamine
$-\overset{\overset{\displaystyle :O:}{\|}}{C}-H$	aldehyde	-al	(structure)	ethanal (acetaldehyde)

(continued)

Common Organic Functional Groups, *(continued)*

Functional Group	Compound Type	Suffix or Prefix of Name	Example	Systematic Name (Common Name)
:O: ‖ —C—C—C—	ketone	-one	H :O: H H—C—C—C—H H H	2-propanone (acetone)
:O: ‖ —C—Ö—H	carboxylic acid	-oic acid	H :O: H—C—C—Ö—H H	ethanoic acid (acetic acid)
:O: ‖ —C—Ö—C—	ester	-oate	H :O: H H—C—C—Ö—C—H H H	methyl ethanoate (methyl acetate)
:O: ‖ —C—N̈—	amide	-amide	H :O: H—C—C—N̈—H H H	ethanamide (acetamide)
—C≡N:	nitrile	-nitrile	H H—C—C≡N: H	ethanenitrile (acetonitrile, methyl cyanide)

The polymerization of chloroethylene, CH_2CHCl (aka vinyl chloride), takes place by the splitting of one of the bonds of the double bond to form a highly reactive carbon molecule containing a single electron at either end of the molecule.

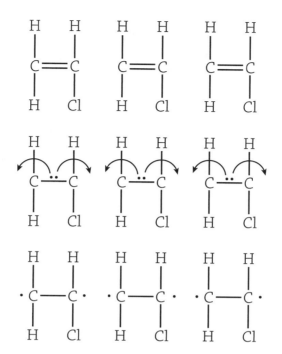

Each end electron then pairs with another end electron on a different molecule, forming a chain.

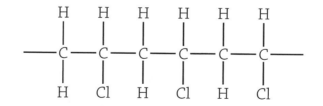

The common name of this polymer is polyvinyl chloride or PVC.

Proteins

Proteins are naturally occurring organic polymers that are composed of monomer units called amino acids. Amino acids contain a carbon atom with two functional groups, an amino group, ($-NH_2$), and a carboxylic acid group, ($-COOH$), attached. This central carbon atom also has a hydrogen atom and another organic group, an "R" group, attached. The identity of the R group determines the identity of the amino acid.

The amino acids link together by the reaction of an amino group of one amino acid with the carboxylic acid group of another amino acid. This forms a peptide bond. (For the sake of clarity, the + signs in the reaction have been omitted.) A peptide bond is an amide group in molecules that are not biochemical. Many functional groups have different names in organic chemistry and in biochemistry.

There are 20 amino acids found in the human body that are involved in protein synthesis. Refer to the internet for a table of the 20 amino acids. The sequence of the amino acids dictates the properties of a protein. Examples of proteins include keratin in hair, hemoglobin, insulin, antibodies, and enzymes.

Carbohydrates

Carbohydrates are made entirely of carbon, hydrogen, and oxygen. The simplest class of carbohydrates is the monosaccharides. Glucose, $C_6H_{12}O_6$, is an example of a monosaccharide. We can make a disaccharide by joining two monosaccharides. Examples of disaccharides include sucrose and lactose. We can make a polysaccharide (polymer), such as starch or cellulose, by joining large numbers of monosaccharide units together.

The monosaccharides have a couple of characteristics that prove to be important in terms of their structure and function. They all have at least one carbon that is bonded to four different groups (a chiral carbon), and most form five- and six-membered rings easily. The presence of chiral carbons allow these compounds to exist as two different optical isomers that are nonsuperimposable mirror images of each other. (Your feet are nonsuperimposable mirror images. Try putting a left shoe on a right foot!)

Glucose may form a chain type structure such as the one that follows. All the carbon atoms except the one at the top and the one at the bottom are chiral. The four different groups making the fourth carbon chiral are outlined.

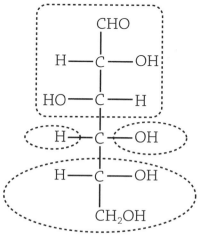

In solution, the chain structure of glucose is in equilibrium with two different ring forms of glucose. The two different ring structures of glucose are α-glucose and β-glucose. These two forms differ in the position of an −OH group relative to the carbon at the far-right side of the ring.

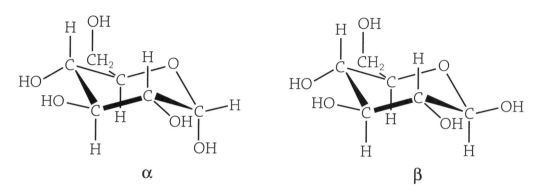

Polysaccharides formed using these two different types of glucose will have different properties. Joining units of β-glucose results in the formation of cellulose. However, if we join units of α-glucose, the result is starch. A minor difference in structural linkage makes the difference between a tree (cellulose) and a potato (starch). This difference may also be seen when you try to eat/digest a tree or a potato.

Nucleic Acids

Nucleic acids are the molecules in our cells that direct and store information for reproduction and cellular growth. There are two types of nucleic acids: ribonucleic acid (RNA) and deoxyribonucleic acid (DNA). Both nucleic acids are unbranched organic polymers composed of monomer units called nucleotides. These nucleotides are composed of a sugar molecule, a nitrogen base, and phosphoric acid. A single DNA molecule may contain several million of these nucleotides, while the smaller RNA molecules may contain several thousand.

The DNA carries the genetic information for the cells. Sections of a DNA molecule called genes contain the information to make a protein. DNA serves two main functions. Molecules of DNA can produce other DNA molecules and RNA molecules. RNA molecules are directly responsible for the synthesis of proteins.

Organic Reactions Problems

There are far too many organic reactions for us to discuss all of them here. We will concentrate on a few important types and give some guidelines on what to look for in other cases. In most cases, the key to any organic reaction is what functional group(s) is/are present. Typically, the functional groups will change and nothing else. A functional group is anything other than a C-C single bond or a C-H bond.

All organic compounds will undergo combustion reactions. That is, they will combine with oxygen gas, O_2, to produce carbon dioxide, CO_2, and water, H_2O. Nitrogen will yield nitrogen gas, N_2. Most of the halogens will produce the appropriate hydrogen halide. For example, chlorine, Cl, gives hydrogen chloride, HCl. The other elements usually give their oxides.

The alkanes are the only organic compounds with no functional group. For this reason, they do not react readily other than by combustion. However, one of the few types of reactions in which the alkanes participate are substitution reactions. As the name implies, something will substitute for something else in the alkane. Normally the reaction involves an alkane, a halogen (either chlorine or bromine), and light. Light is necessary to initiate the reaction. In a substitution reaction of this type, a halogen atom substitutes for a hydrogen reaction. This process may continue until halogen atoms replace all the hydrogen atoms. To minimize the opportunity for multiple replacements occurring, an excess of alkane is normally present. The following is an example of a substitution reaction. The symbol, $h\nu$, indicates light energy.

If combustion occurs with a slight deficiency of oxygen, then carbon monoxide, CO, may form instead of carbon dioxide, CO_2. A severe deficiency of oxygen may result in elemental carbon, C, forming instead of carbon dioxide or carbon monoxide.

$$CH_3CH_3 + Cl_2 \xrightarrow{h\nu} CH_3CH_2Cl + HCl$$

Aromatic hydrocarbons, like alkanes, undergo substitution reactions. The conditions required for an aromatic compound to react are different than that for the reaction of an alkane. Aromatic substitution reactions are beyond the scope of this text.

Alkenes and alkynes typically react by addition reactions. As the name implies, two or more molecules simply add together. Normally, one product results with a formula that is simply the sum of the reactant pieces. Only the functional group, the carbon-carbon double or triple bond, will change during the reaction.

In many organic reactions, compounds such as HCl are not included in the final answer. The key to the organic reaction equation is what happens to the organic compound and not perfecting a balanced chemical equation.

Two typical addition reactions are hydrogenation and halogenation. In hydrogenation, we add a hydrogen, H_2, molecule. Thus, C_2H_4 will become $C_2H_4 + H_2 = C_2H_6$. Hydrogenation always requires a catalyst. The most common catalysts are platinum, Pt, palladium, Pd, or nickel, Ni. In a halogenation reaction, we add a halogen, either chlorine or bromine. No catalyst is necessary, and unlike the reaction with an alkane, light is not necessary. The formation of a polymer, such as polypropylene, is an addition reaction. For this reason, polypropylene is an example of an addition polymer.

The following illustrates the hydrogenation and halogenation, with bromine, of the alkene propene. Notice that the $-CH_3$ group, which is not a functional group, does not change during either reaction.

$$CH_2{=}CH{-}CH_3 \ + \ H_2 \xrightarrow{\text{catalyst}} \underset{\displaystyle CH_2{-}\underset{\displaystyle |}{\overset{\displaystyle |}{CH}}{-}CH_3}{\overset{\displaystyle \overset{H}{|} \quad \overset{H}{|}}{}}$$

$$CH_2{=}CH{-}CH_3 \ + \ Cl_2 \longrightarrow \underset{\displaystyle CH_2{-}CH{-}CH_3}{\overset{\displaystyle \overset{Cl}{|} \quad \overset{Cl}{|}}{}}$$

The hydrogenation of an alkyne can give two possible products. If there is a limited quantity of hydrogen available, hydrogenation converts an alkyne to an alkene. If more hydrogen is available, the alkene will react further to become an alkane. The following illustrates the hydrogenation of propyne. Only the first step occurs if there is not much hydrogen, while the second step occurs if there is enough hydrogen remaining after the first step.

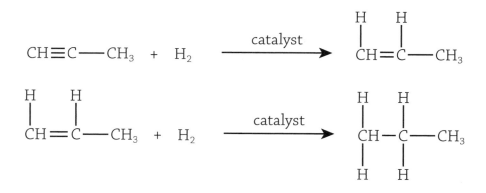

The halogenation of an alkyne, like hydrogenation, can give two possible products. A limited amount of halogen will only add one molecule, whereas an excess of halogen will add two molecules.

The other functional groups participate in a variety of reactions. We have already seen some of these reactions. The carboxylic acids will behave as typical weak acids. The amines will behave as typical weak bases. The only other category of reaction we will examine here is a condensation reaction.

A **condensation reaction** joins two molecules and splits out a small molecule. The small molecule is usually water. The formation of a peptide bond is an example of a condensation reaction, as in the following figure. The conditions necessary for a condensation reaction vary with the functional groups involved. In most cases, a catalyst will be present. The two most common catalysts are acids and enzymes. Two alcohols will condense to form an ether. A carboxylic acid condenses with an alcohol to form an ester. A carboxylic acid condenses with an amine to form an amide.

BTW
A carboxylic acid can react with an amine via an acid-base reaction. Take care to bypass this reaction.

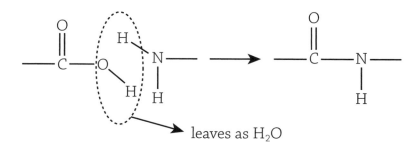

The following illustrates the reaction of two alcohols to form an ether. The reacting molecules are both molecules of α-glucose. This reaction occurs in the presence of an enzyme catalyst, which prevents the other functional groups from reacting. In biochemistry, this ether group is part of a **glycoside linkage**. The joining of two molecules of the monosaccharide, β-glucose, gives the disaccharide, lactose. The joining of additional β-glucose molecules will eventually generate the polysaccharide, cellulose. If we replace the β-glucose with α-glucose, the disaccharide would be maltose. A polysaccharide of α-glucose is starch.

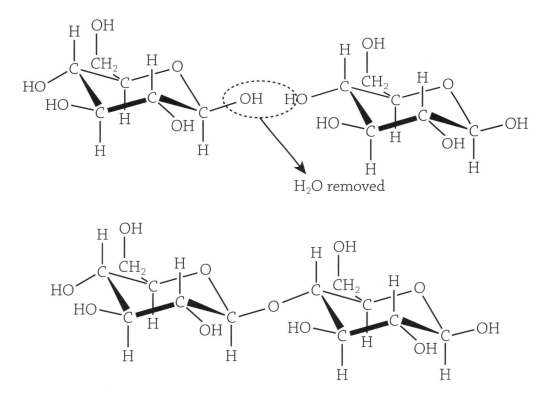

Let's do a problem illustrating how a carboxylic acid reacts with an alcohol to form an ester.

EXAMPLE

▶ Use structural formulas to illustrate how the following molecules react:

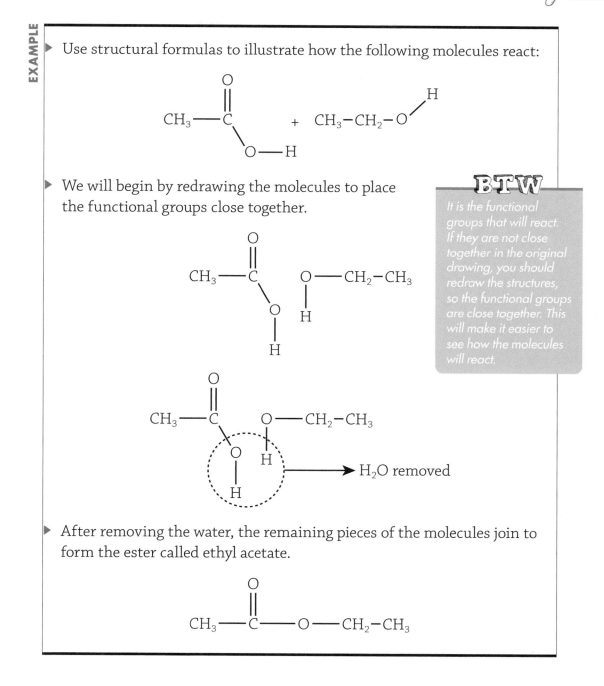

▶ We will begin by redrawing the molecules to place the functional groups close together.

BTW

It is the functional groups that will react. If they are not close together in the original drawing, you should redraw the structures, so the functional groups are close together. This will make it easier to see how the molecules will react.

▶ After removing the water, the remaining pieces of the molecules join to form the ester called ethyl acetate.

It is possible to reverse the formation of an ester by a catalyzed hydrolysis reaction. In this reaction, the water reenters the molecule where it was removed. If an acid is the catalyst, the original acid and alcohol will reform. If a base is the catalyst, the alcohol will reform; however, the acid will react further to produce its conjugate base. The base-catalyzed hydrolysis of an ester is **saponification**.

 IRL Saponification reactions are important in the production of soaps.

If a molecule has multiple functional groups, multiple reactions may occur. For example, the reaction of molecules with two carboxylic acid groups might react with molecules containing two alcohol groups in the following manner.

After multiple condensation reactions, the molecules join to give:

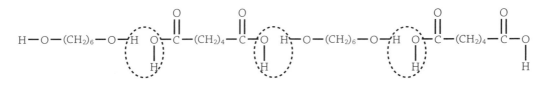

The resultant molecule is a polymer held together by many ester groups. Such a polymer is a polyester. The linkage through multiple condensation reactions results in a **condensation polymer**.

 IRL Some examples of condensation polymers are polyester fabric, nylon, and silicones.

If we replace the di-alcohol in the preceding reaction with a di-amine we get:

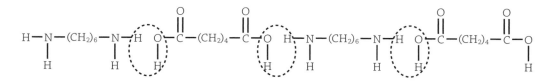

After multiple condensation reactions, the molecules join to give:

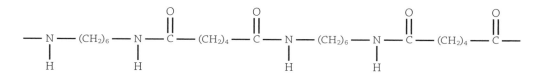

Multiple amide groups link the resultant polymer. For this reason, this condensation polymer is a polyamide; however, DuPont did not like this name, so we now know this polymer to be an example of a nylon. If the units joined by the condensation reactions were amino acids, the amide groups would be peptide bonds, and the polymer would be a polypeptide or a protein.

EXERCISES

EXERCISE 21-1

Answer the following questions.

1. How many carbon atoms may bond together?

2. What are the four types of hydrocarbons?

3. What type(s) of organic compounds react in combustion reactions?

4. What is the general formula of an alkane?

5. How is the general formula of a cycloalkane like the general formula of an alkene?

6. When naming an alkane, what suffix must be present?

7. In organic compounds, how many bonds does carbon always have?

8. When an organic molecule reacts, where is the most likely site of reaction?

9. Write a balanced chemical equation showing how three molecules of ethylene, $CH_2=CH_2$, react to form polyethylene.

10. A polymer of amino acids is _____.

11. Circle each of the chiral carbon atoms in the following molecule.

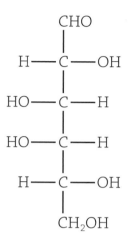

EXERCISE 21-2

Name the following compounds.

1. CH$_3$—CH$_2$
      |
    CH$_2$—CH$_3$

3.
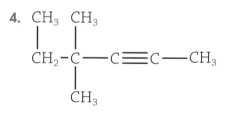

2.
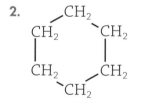

4. CH$_3$ CH$_3$
      | |
    CH$_2$—C—C≡C—CH$_3$
        |
       CH$_3$

Flashcard App

Answer Key

1

Getting Started in Chemical Calculations

EXERCISE 1-1

1. These are the properties of a solid.

2. μ is the abbreviation for *micro-*, which is 10^{-6} or 0.000001.

3. The answer will have 2 significant figures because of the 4.5 in the calculation.

4. The "?" must match the unit listed, which is feet.

5. **False**   Liquids do not have a variable volume. The description is that of a gas.

6. **False**   *M* is the abbreviation for *mega-*. *Milli-* has the abbreviation *m*.

7. **True**   The values sum to 12.0 (all the values used in the problem have one decimal place; therefore, the answer must have one decimal place).

8. **True**   Cubing the 12 gives 1728.

9. **b**   The other answers involve mathematical errors.

10. **d**   You are given 1 kg = 1000 g, which appears only in C and D. You are also given 1 cm = 0.01 m, which appears only in A and D.

11. $(15.2 \text{ in})\left(\dfrac{2.54 \text{ cm}}{1 \text{ in}}\right) = 38.6 \text{ cm}$ This is the shortest way to make this conversion.

12. $(15.2 \text{ cm})\left(\dfrac{1 \text{ in}}{2.54 \text{ cm}}\right) = 5.98 \text{ in}$ This is the shortest way to make this conversion.

13. $(2.5 \text{ ft}^3)\left(\dfrac{12 \text{ in}}{1 \text{ ft}}\right)^3 = 4.3 \times 10^3 \text{ in}^3$ You are beginning with a cubed unit; therefore, the conversion needs to be cubed.

14. $(2.53 \text{ lb})\left(\dfrac{453.59 \text{ g}}{1 \text{ lb}}\right) = 1.15 \times 10^3 \text{ g}$ This is the shortest way to make this conversion.

15. $(2.0 \text{ L})\left(\dfrac{1.057 \text{ qt}}{1 \text{ L}}\right) = 2.1 \text{ qt}$ This is the shortest way to make this conversion.

16. $(1.00 \text{ qt})\left(\dfrac{1 \text{ L}}{1.057 \text{ qt}}\right)\left(\dfrac{1 \text{ mL}}{0.001 \text{ L}}\right) = 946 \text{ mL}$ You could save a step if you knew that 946.1 mL = 1 qt.

17. $(2.205 \text{ lb})\left(\dfrac{453.59 \text{ g}}{1 \text{ lb}}\right)\left(\dfrac{1 \text{ kg}}{10^3 \text{ g}}\right) = 1.000 \text{ kg}$ If you knew the number of pounds in a kilogram, you could save one step, but why waste brain cells when you can get the answer this way.

18. $(4.00 \text{ yd}^2)\left(\dfrac{3 \text{ ft}}{1 \text{ yd}}\right)^2\left(\dfrac{30.48 \text{ cm}}{1 \text{ ft}}\right)^2 = 3.34 \times 10^4 \text{ cm}^2$ The initial value is a squared unit, so each of the conversions also needs to be squared.

19. $(3.27 \text{ t})\left(\dfrac{2000 \text{ lb}}{1 \text{ t}}\right)\left(\dfrac{453.59 \text{ g}}{1 \text{ lb}}\right)\left(\dfrac{1 \text{ kg}}{10^3 \text{ g}}\right) = 2.97 \times 10^3 \text{ kg}$ This problem assumes you already know the US conversion that 1 ton = 2000 lbs.

20. $(2.0 \text{ m}^2)\left(\dfrac{1 \text{ cm}}{0.01 \text{ m}}\right)^2\left(\dfrac{1 \text{ in}}{2.54 \text{ cm}}\right)^2\left(\dfrac{1 \text{ ft}}{12 \text{ in}}\right)^2 = 22 \text{ ft}^2$ Do not forget that a zero after the decimal point is a significant figure.

21. $\left(\dfrac{52\text{ g}}{15\text{ mL}}\right) = 3.5\text{ g/mL}$ Simply enter the given values into the definition of density.

22. $\left(15\text{ cm}^3\right)\left(\dfrac{0.8255\text{ g}}{\text{cm}^3}\right) = 12\text{ g}$ This is simpler than treating the definition of density as a mathematical relationship.

23. $\left(17.5\text{ g}\right)\left(\dfrac{\text{mL}}{0.7826\text{ g}}\right) = 22.4\text{ mL}$ Note, to get the units correct, it is

necessary to invert the density. This can be done for any relationship when necessary; for example, these are both valid conversions:
$\left(\dfrac{12\text{ in}}{1\text{ ft}}\right)$ and $\left(\dfrac{1\text{ ft}}{12\text{ in}}\right)$.

24. $\left(14.2\text{ kg}\right)\left(\dfrac{10^3\text{ g}}{1\text{ kg}}\right)\left(\dfrac{\text{cm}^3}{7.8\text{ g}}\right)\left(\dfrac{0.01\text{ m}}{1\text{ cm}}\right)^3 = 1.8 \times 10^{-3}\text{ m}^3$ Note, these

steps may be done in any order and, if desired, other conversions could be used.

25. $\left(\dfrac{125\text{ lb}}{2.00\text{ ft}^3}\right)\left(\dfrac{453.59\text{ g}}{1\text{ lb}}\right)\left(\dfrac{1\text{ ft}}{12\text{ in}}\right)^3\left(\dfrac{1\text{ in}}{2.54\text{ cm}}\right)^3 = 1.00\text{ g/cm}^3$ Note, these

steps may be done in any order, and if desired, other conversions could be used.

Atoms, Ions, and Molecules

EXERCISE 2-1

1. **neutron**   slightly heavier than a proton

2. **electron**   Only protons and neutrons are inside the nucleus.

3. **metals** except hydrogen

4. **periods** Columns are the groups or families.

5. **cations** Anions have a negative charge.

6. **True** It is slightly heavier.

7. **False** The atomic mass appears, which is not the same as the mass number.

8. **False** There are other differences, such as how they conduct electricity.

9. **False** Compounds never have a charge, only ions, protons, and electrons have charges.

10. **a** The alkali metals and hydrogen normally have a +1 charge in their compounds.

11. **c** The alkaline earth metals normally have a +2 charge in their compounds.

12. **b** The halogens normally have a –1 charge in their compounds.

13. **d** The elements in the oxygen family normally have a –2 charge in their compounds.

14. **a** Na is an alkali metal and alkali metals normally form +1 ions. In addition, Cl is –1.

15. **d** Ca is an alkaline earth metal and alkaline earth metals normally form +2 ions. In addition, O is –2.

16. **f** Elements in an uncombined state have 0 charge.

17. **e** The name is iron(III) phosphide. The (III) refers to a +3 charge on the iron.

18. **d** One Mn contributes +4, and it is necessary to have a –4 charge to balance it. It takes two oxide ions (–2 each) to total –4.

19. **c** The prefix *di-* = 2 (Cl_2) and the prefix *tri-* = 3 (O_3).

20.

	$^{23}_{11}\text{Na}$	$^{40}_{19}\text{K}^+$	$^{53}_{25}\text{Mn}^{2+}$	$^{130}_{54}\text{Xe}$	$^{75}_{33}\text{As}^{3-}$
Protons	11	19	25	54	33
Neutrons	12	21	28	76	42
Electrons	11	18	23	54	36
Charge	0	+1	+2	0	−3
Mass number	23	40	53	130	75

The protons give the atomic number, which is the lower left subscript in the chemical symbol. The electrons equal the protons with an adjustment for the charge: add an electron for every negative charge and subtract an electron for every positive charge. The neutrons are the mass number minus the protons. The charge is the protons minus the electrons. It is also the number to the upper right of the chemical symbol. The mass number is the number of protons plus the number of neutrons.

21. These are metal–nonmetal compounds, so multiplying prefixes, such as *di-*, are not used. The name of the second element in the formula must have an *-ide* suffix.

    a. sodium sulfide: Na = sodium and S = sulfide

    b. magnesium chloride: Mg = magnesium and Cl = chloride

    c. aluminum fluoride: Al = aluminum and F = fluoride

    d. copper(II) oxide: Cu = copper and O = oxide   Since the Cu is a transition metal, it is necessary to designate its charge. In this case, the charge on Cu is +2 (to cancel −2 from O), which leads to copper(II).

    e. potassium hydride: K = potassium and H = hydride

22.   a. $CaCl_2$: calcium = Ca and chloride = Cl^-

     b. Na_2O: sodium = Na and oxide = O^{2-}

     c. AlN: aluminum = Al and nitride = N^{3-}

     d. MnF_2: manganese(II) = Mn^{2+} and fluoride = F^-

     e. MnF_3: manganese(III) = Mn^{3+} and fluoride = F^-

23. These are compounds containing only nonmetals, so prefixes are necessary (except for hydrogen). Prefixes ending in -o or -a may have the terminal vowel dropped if the name of the element begins with a vowel

     a. nitrogen dioxide: N = nitrogen, O = oxide, 2 = di-

     b. carbon monoxide: C = carbon, O = oxide, 1 = mono- (one of the few places where this prefix is still used)

     c. hydrogen sulfide or hydrosulfuric acid: H = hydrogen, S = sulfide, no prefix for H, hydrosulfuric acid is an alternate name, used because many H compounds are acids

     d. dichlorine pentoxide: Cl = chlorine, 2 = di-, O = oxide, 5 = penta-

     e. methane: this is a special name such as water

24.   a. HCl: hydrogen = H and chloride = Cl

     b. SO_2: sulfur = S and oxide = O, di- = 2

     c. SO_3: sulfur = S and oxide = O, tri- = 2

     d. N_2O_5: nitrogen = N, di- = 2, oxide = O, and pent- = 5

     e. NH_3: this is a special name such as water

25.   a. sodium nitrate: Na^+ = sodium ion and NO_3^- = nitrate ion (charges balance)

     b. potassium carbonate: K^+ = potassium ion and CO_3^{2-} = carbonate ion (balance charges)

     c. calcium sulfate: Ca^{2+} = calcium ion and SO_4^{2-} = sulfate ion (charges balance)

     d. ammonium phosphate: NH_4^+ = ammonium ion and PO_4^{3-} = phosphate ion (balance charges)

     e. iron(II) nitrite: Fe^{2+} = iron(II) ion and NO_2^- = nitrite ion (balance charges)

26. a. Na_3PO_4: sodium ion = Na^+ and phosphate ion = PO_4^{3-} (balance charges)

  b. NH_4HCO_3: ammonium ion = NH_4^+ and bicarbonate ion = HCO_3^- (balance charges)

  c. $Mg(NO_2)_2$: magnesium ion = Mg^{2+} and nitrite ion = NO_2^- (balance charges)

  d. CuCl: copper(I) ion = Cu^+ and chloride ion = Cl^- (charges balance)

  e. Fe_2O_3: iron(III) = Fe^{3+} and oxide = O^{2-} (balance charges)

27. The names of acids must include the name *acid*.

  a. hydrofluoric acid: a binary acid

  b. nitric acid: a ternary acid containing the nitrate ion

  c. phosphoric acid: a ternary acid containing the phosphate ion

  d. carbonic acid: a ternary acid containing the carbonate ion

  e. sulfuric acid: a ternary acid containing the sulfate ion

28. Acids with a *hydro-* prefix are binary acids all others on this list are ternary acids

  a. HNO_2: contains the nitrite ion

  b. H_2S: contains the sulfide ion

  c. H_2SO_4: contains the sulfate ion

  d. HCl: contains the chloride ion

  e. HNO_3: contains the nitrate ion

29. Locate the element on the periodic table (do not forget that hydrogen is an exception).

  a. metal

  b. metal

  c. metal

  d. nonmetal

  e. nonmetal

  f. nonmetal

30. Locate the element on the periodic table and note which column the element is in. Metals give cations and nonmetals give anions.

   a. Cs^+
   b. Ra^{2+}
   c. Tl^{3+}
   d. P^{3-}
   e. Se^{2-}
   f. I^-

Moles, Stoichiometry, and Equations

EXERCISE 3-1

1. reactants (the products are on the right side)

2. law of conservation of mass

3. Avogadro's number, 6.022×10^{23}

4. molar mass

5. 28 g/mol (= 12 g/mol + 16 g/mol)

6. limiting reactant (reagent)

7. theoretical yield

8. empirical formula

9.
   a. HO   divide both subscripts by 2
   b. CH_2O divide all subscripts by 6
   c. CH_3 divide all subscripts by 2
   d. HNO_3 no further simplification is possible
   e. $(NH_4)SO_4$ divide all subscripts, except NH_4, by 2

10. Begin by determining the moles of each element:

$$(21.4 \text{ g S}) \left(\frac{1 \text{ mol S}}{32.07 \text{ g S}} \right) = 0.667 \text{ mol S}$$

$$(50.8 \text{ g F}) \left(\frac{1 \text{ mol F}}{19.0 \text{ g F}} \right) = 2.67 \text{ mol F}$$

Divide each of the moles by the smaller value (0.667):

    0.667 mol S/0.667 = 1 S

    2.67 mol F / 0.667 = 4 F

The results give you: empirical formula is SF_4.

11. Begin by determining the moles of each element (using the hint):

$$(69.6 \text{ g Mn}) \left(\frac{1 \text{ mol Mn}}{54.94 \text{ g Mn}} \right) = 1.27 \text{ mol Mn}$$

$$(30.4 \text{ g O}) \left(\frac{1 \text{ mol O}}{16.0 \text{ g O}} \right) = 1.90 \text{ mol O}$$

Divide each of the moles by the smaller value (1.27):

    1.27 mol Mn/1.27 = 1 Mn

    1.90 mol O/1.27 = 1.50

The oxygen value is too far from a whole number to round; therefore, multiply all values by the smallest value to get a whole number (× 2): empirical formula is Mn_2O_3.

12. It is important to determine the empirical formula first. Start by determining the moles of each element:

$$(25.2 \text{ g S}) \left(\frac{1 \text{ mol S}}{32.07 \text{ g S}} \right) = 0.786 \text{ mol S}$$

$$(74.8 \text{ g F}) \left(\frac{1 \text{ mol F}}{19.0 \text{ g F}} \right) = 3.94 \text{ mol F}$$

Divide each of the moles by the smaller value (0.786):

0.786 mol S/0.786 = 1 S   3.94 mol F/0.786 = 5 F

The results give you: Empirical formula is SF_5. Finish the problem by converting the empirical formula to the molecular formula. This is done by comparing the molar masses. The molar mass of the empirical formula is 127.01 g/mol, which is half the molar mass given in the problem. This means the empirical formula is one-half the molecular formula. Therefore, the molecular formula is S_2F_{10}.

EXERCISE 3-2

1. $2 N_2(g) + 5 O_2(g) \rightarrow 2 N_2O_5(s)$
2. $2 CoCl_2(s) + 2 ClF_3(g) \rightarrow 2 CoF_3(s) + 3 Cl_2(g)$
3. $2 La(OH)_3(s) + 3 H_2C_2O_4(aq) \rightarrow La_2(C_2O_4)_3(s) + 6 H_2O(l)$
4. $2 C_4H_{10}(g) + 13 O_2(g) \rightarrow 8 CO_2(g) + 10 H_2O(g)$
5. $2 NH_3(g) + 3 F_2(g) \rightarrow N_2(g) + 6 HF(g)$

EXERCISE 3-3

The mole ratios will be the coefficient of first (only) product over the moles of the second reactant.

1. $\left(3.00 \text{ mol } O_2\right)\left(\dfrac{2 \text{ mol } N_2O_5}{5 \text{ mol } O_2}\right) = 1.20 \text{ mol } N_2O_5$

2. $\left(3.00 \text{ mol } ClF_3\right)\left(\dfrac{2 \text{ mol } CoF_3}{2 \text{ mol } ClF_3}\right) = 3.00 \text{ mol } CoF_3$

3. $\left(3.00 \text{ mol } H_2C_2O_4\right)\left(\dfrac{1 \text{ mol } La_2(C_2O_4)_3}{3 \text{ mol } H_2C_2O_4}\right) = 1.00 \text{ mol } La_2(C_2O_4)_3$

4. $\left(3.00 \text{ mol } O_2\right)\left(\dfrac{8 \text{ mol } CO_2}{13 \text{ mol } O_2}\right) = 1.85 \text{ mol } CO_2$

5. $\left(3.00 \text{ mol } F_2\right)\left(\dfrac{1 \text{ mol } N_2}{3 \text{ mol } F_2}\right) = 1.00 \text{ mol } N_2$

EXERCISE 3-4

1. $2 \text{ Br}_2(l) + 3 \text{ O}_2(g) \rightarrow 2 \text{ Br}_2\text{O}_3(s)$
2. $\text{PF}_3(l) + 3 \text{ H}_2\text{O}(l) \rightarrow \text{H}_3\text{PO}_3(aq) + 3 \text{ HF}(aq)$
3. $3 \text{ Zn(OH)}_2(s) + 2 \text{ H}_3\text{VO}_4(aq) \rightarrow \text{Zn}_3(\text{VO}_4)_2(s) + 6 \text{ H}_2\text{O}(l)$
4. $2 \text{ C}_6\text{H}_{14}(l) + 19 \text{ O}_2(g) \rightarrow 12 \text{ CO}_2(g) + 14 \text{ H}_2\text{O}(l)$
5. $\text{CH}_4(g) + 4 \text{ Cl}_2(g) \rightarrow \text{CCl}_4(l) + 4 \text{ HCl}(g)$

EXERCISE 3-5

In each case, divide the moles of the reactant (3.00 mol) by the coefficient from in the reaction, and the smaller number is limiting.

1. We have 3.00 mol $\text{Br}_2/2 = 1.50$ and 3.00 mol $\text{O}_2/3 = 1.00$, thus the O_2 is limiting.

2. We have 3.00 mol $\text{PF}_3/1 = 3.00$ and 3.00 mol $\text{H}_2\text{O}/3 = 1.00$, thus the H_2O is limiting.

3. We have 3.00 mol $\text{Zn(OH)}_2/3 = 1.00$ and 3.00 mol $\text{H}_3\text{VO}_4/2 = 1.50$, thus the Zn(OH)_2 is limiting.

4. We have 3.00 mol $\text{C}_6\text{H}_{14}/2 = 1.50$ and 3.00 mol $\text{O}_2/19 = 0.158$, thus the O_2 is limiting.

5. We have 3.00 mol $\text{CH}_4/1 = 13.00$ and 3.00 mol $\text{Cl}_2/4 = 0.750$, thus the Cl_2 is limiting.

EXERCISE 3-6

Using the limiting reactant from 3-5, it is possible to determine the theoretical yield. The percent yield is $= \dfrac{\text{actual yield}}{\text{theoretical yield}} \times 100\%$, with the actual yield being 10.0 g.

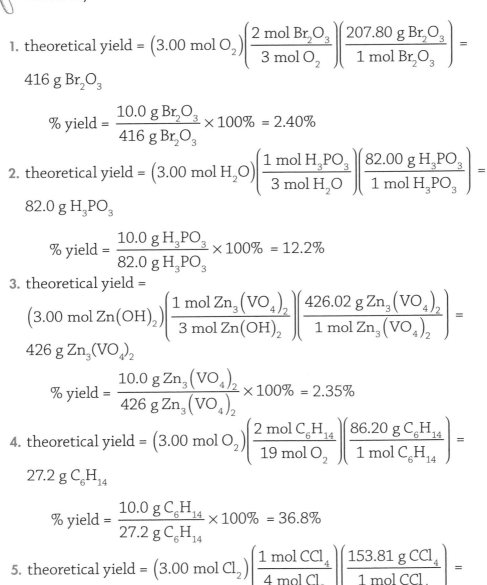

1. theoretical yield $= \left(3.00 \text{ mol } O_2\right)\left(\dfrac{2 \text{ mol } Br_2O_3}{3 \text{ mol } O_2}\right)\left(\dfrac{207.80 \text{ g } Br_2O_3}{1 \text{ mol } Br_2O_3}\right) =$

416 g Br_2O_3

% yield $= \dfrac{10.0 \text{ g } Br_2O_3}{416 \text{ g } Br_2O_3} \times 100\% = 2.40\%$

2. theoretical yield $= \left(3.00 \text{ mol } H_2O\right)\left(\dfrac{1 \text{ mol } H_3PO_3}{3 \text{ mol } H_2O}\right)\left(\dfrac{82.00 \text{ g } H_3PO_3}{1 \text{ mol } H_3PO_3}\right) =$

82.0 g H_3PO_3

% yield $= \dfrac{10.0 \text{ g } H_3PO_3}{82.0 \text{ g } H_3PO_3} \times 100\% = 12.2\%$

3. theoretical yield $=$

$\left(3.00 \text{ mol } Zn(OH)_2\right)\left(\dfrac{1 \text{ mol } Zn_3(VO_4)_2}{3 \text{ mol } Zn(OH)_2}\right)\left(\dfrac{426.02 \text{ g } Zn_3(VO_4)_2}{1 \text{ mol } Zn_3(VO_4)_2}\right) =$

426 g $Zn_3(VO_4)_2$

% yield $= \dfrac{10.0 \text{ g } Zn_3(VO_4)_2}{426 \text{ g } Zn_3(VO_4)_2} \times 100\% = 2.35\%$

4. theoretical yield $= \left(3.00 \text{ mol } O_2\right)\left(\dfrac{2 \text{ mol } C_6H_{14}}{19 \text{ mol } O_2}\right)\left(\dfrac{86.20 \text{ g } C_6H_{14}}{1 \text{ mol } C_6H_{14}}\right) =$

27.2 g C_6H_{14}

% yield $= \dfrac{10.0 \text{ g } C_6H_{14}}{27.2 \text{ g } C_6H_{14}} \times 100\% = 36.8\%$

5. theoretical yield $= \left(3.00 \text{ mol } Cl_2\right)\left(\dfrac{1 \text{ mol } CCl_4}{4 \text{ mol } Cl_2}\right)\left(\dfrac{153.81 \text{ g } CCl_4}{1 \text{ mol } CCl_4}\right) =$

115 g CCl_4

% yield $= \dfrac{10.0 \text{ g } CCl_4}{115 \text{ g } CCl_4} \times 100\% = 8.70\%$

EXERCISE 3-7

1. 2 (H) + 1 (O) = 2 (1.01 g/mol) + 1 (16.00 g/mol) = 18.02 g/mol

2. 1 (C) + 2 (O) = 12.01 g/mol + 2 (16.00 g/mol) = 44.01 g/mol

3. 1 (H) + 1 (N) + 3 (O) = (1.01 g/mol) + (14.01 g/mol) + 3 (16.00) = 63.02 g/mol

4. 2 (Na) + (S) + 4 (O) = 2 (22.99) + (32.06) + 4 (16.00) = 142.04 g/mol

5. 3 (N) + 12 (H) + (P) + 4 (O) = 3 (14.01) + 12 (1.01) + (30.97) + 4 (16.00) = 149.12 g/mol

EXERCISE 3-8

In each case, use the molar mass from Exercise 3-7 (it needs to be inverted for the units to cancel).

1. $\left(100.00 \text{ g } H_2O\right)\left(\dfrac{1 \text{ mol } H_2O}{18.02 \text{ g } H_2O}\right) = 5.549 \text{ mol } H_2O$

2. $\left(100.00 \text{ g } CO_2\right)\left(\dfrac{1 \text{ mol } CO_2}{44.01 \text{ g } CO_2}\right) = 2.272 \text{ mol } CO_2$

3. $\left(100.00 \text{ g } HNO_3\right)\left(\dfrac{1 \text{ mol } HNO_3}{63.02 \text{ g } HNO_3}\right) = 1.587 \text{ mol } HNO_3$

4. $\left(100.00 \text{ g } Na_2SO_4\right)\left(\dfrac{1 \text{ mol } Na_2SO_4}{142.04 \text{ g } Na_2SO_4}\right) = 0.70403 \text{ mol } Na_2SO_4$

5. $\left(100.00 \text{ g } \left(NH_4\right)_3 PO_4\right)\left(\dfrac{1 \text{ mol } \left(NH_4\right)_3 PO_4}{149.12 \text{ g } \left(NH_4\right)_3 PO_4}\right) = 0.67060 \text{ mol }$ $(NH_4)_3PO_4$

EXERCISE 3-9

In each case, use the molar mass from Exercise 3-7.

1. $\left(2.50 \text{ mol H}_2\text{O}\right)\left(\dfrac{18.02 \text{ g H}_2\text{O}}{1 \text{ mol H}_2\text{O}}\right) = 45.0 \text{ g mol H}_2\text{O}$

2. $\left(2.50 \text{ mol CO}_2\right)\left(\dfrac{44.01 \text{ g CO}_2}{1 \text{ mol CO}_2}\right) = 110. \text{ g CO}_2$

3. $\left(2.50 \text{ mol HNO}_3\right)\left(\dfrac{63.02 \text{ g HNO}_3}{1 \text{ mol HNO}_3}\right) = 158 \text{ g HNO}_3$

4. $\left(2.50 \text{ mol Na}_2\text{SO}_4\right)\left(\dfrac{142.04 \text{ g Na}_2\text{SO}_4}{1 \text{ mol Na}_2\text{SO}_4}\right) = 355 \text{ g Na}_2\text{SO}_4$

5. $\left(2.50 \text{ mol } (\text{NH}_4)_3\text{PO}_4\right)\left(\dfrac{149.12 \text{ g } (\text{NH}_4)_3\text{PO}_4}{1 \text{ mol } (\text{NH}_4)_3\text{PO}_4}\right) = 373 \text{ g } (\text{NH}_4)_3\text{PO}_4$

Aqueous Solutions

EXERCISE 4-1

1. **False**   A solution is a homogeneous mixture.

2. **True**   All acids are electrolytes; some acids are strong electrolytes, and some are weak electrolytes.

3. **c**   All the other compounds contain a metal and a nonmetal or at least one polyatomic ion.

4. **a**   All the compounds have Cl⁻, and AgCl is listed as one of the insoluble chlorides.

5. **e** $BaSO_4$ is one of the four sulfates listed as being insoluble.

6. **d**   The possible products are:
   a. $AgBr$ (insoluble) and KNO_3 (soluble)
   b. $PbCl_2$ (insoluble) and KNO_3 (soluble)
   c. $NaCl$ (soluble) and $Fe(OH)_2$ (insoluble)
   d. $(NH_4)_3PO_4$ (soluble) and $NaCl$ (soluble)
   e. $CaSO_4$ (insoluble) and KCl (soluble)

7. **b**   The strong acids are: HCl, HBr, HI, HNO_3, H_2SO_4, $HClO_3$, and $HClO_4$

8. **acids**   Nonmetal oxides are usually acid anhydrides.

9. **bases**   Metal oxides are often base anhydrides.

10. **acid, base**   All neutralization reactions involve an acid and a base.

11. **base, acid**   The cation is always from the base and the anion is always from the acid.

12. **redox**   The abbreviation comes from reduction and oxidation.

13. **strong**   The strong electrolytes are strong acids, strong bases, and soluble ionic compounds.

14. **spectator**   The spectator ions are any ions appearing on both sides of the reaction arrow.

15. **endpoint**   The theoretical end of a titration is the equivalent point.

EXERCISE 4-2

1. **Mg**   Mg atoms are losing electrons to become Mg ions.

2. **Ag** Ag^+ are gaining electrons to become Ag metal.

3. **Ag** Ag^+ caused the oxidation of Mg.

4. **Mg**   Mg caused the reduction of Ag^+.

5. **0**   The charge of any element is 0.

EXERCISE 4-3

In each case, begin by having all the ions "change partners." If anything that might form is a weak electrolyte, a nonelectrolyte, or a precipitate, a reaction will occur. If all possible products are strong electrolytes, the answer is NR.

1. $Ba(OH)_2(aq) + H_2SO_4(aq) \rightarrow BaSO_4(s) + 2\ H_2O(l)$
2. $2\ KOH(aq) + FeCl_2(aq) \rightarrow Fe(OH)_2(s) + 2\ KCl(aq)$
3. $NH_4NO_3(aq) + Na_2SO_4(aq) \rightarrow NR$
4. $2\ HC_2H_3O_2(aq) + CaCO_3(s) \rightarrow Ca(C_2H_3O_2)_2(aq) + H_2O(l) + CO_2(g)$
5. $2\ K_3PO_4(aq) + 3\ Ca(NO_2)_2(aq) \rightarrow Ca_3(PO_4)_2(s) + 6\ KNO_2(aq)$

EXERCISE 4-4

Separate all strong electrolytes to the appropriate ions (do not forget the take coefficients and subscripts into account). Do not alter the formulas of any weak or nonelectrolyte.

1. $Ba^{2+}(aq) + 2\ OH^-(aq) + 2\ H^+(aq) + SO_4^{2-}(aq) \rightarrow BaSO_4(s) + 2\ H_2O(l)$
2. $2\ K^+(aq) + 2\ OH^-(aq) + Fe^{2+}(aq) + 2\ Cl^-(aq) \rightarrow Fe(OH)_2(s) + 2\ K^+(aq) + 2\ Cl^-(aq)$
3. No reaction, but you can write: $NH_4^+(aq) + NO_3^-(aq) + 2\ Na^+(aq) + SO_4^{2-}(aq) \rightarrow NR$
4. $2\ HC_2H_3O_2(aq) + CaCO_3(s) \rightarrow Ca^{2+}(aq) + 2\ C_2H_3O_2^-(aq) + H_2O(l) + CO_2(g)$
5. $6\ K^+(aq) + 2\ PO_4^{3-}(aq) + 3\ Ca^{2+}(aq) + 6\ NO_2^-(aq) \rightarrow Ca_3(PO_4)_2(s) + 6\ K^+(aq) + 6\ NO_2^-(aq)$

EXERCISE 4-5

Begin with the answers (ionic equations) to exercise 4-4. Remove any ions appearing on each side of the reaction arrow in the ionic equations (the formulas must be an exact match).

1. $Ba^{2+}(aq) + 2\ OH^-(aq) + 2\ H^+(aq) + SO_4^{2-}(aq) \rightarrow BaSO_4(s) + 2\ H_2O(l)$
2. $2\ OH^-(aq) + Fe^{2+}(aq) \rightarrow Fe(OH)_2(s)$

3. no reaction

4. $2\ HC_2H_3O_2(aq) + CaCO_3(s) \rightarrow Ca^{2+}(aq) + 2\ C_2H_3O_2^{-}(aq) + H_2O(l) + CO_2(g)$

5. $2\ PO_4^{3-}(aq) + 3\ Ca^{2+}(aq) \rightarrow Ca_3(PO_4)_2(s)$

EXERCISE 4-6

Find the reactants on the activity series. If the element (no charge) is above the ion, there will be a reaction. If there is a reaction, each reactant will move to the opposite side of the arrow in the activity series.

1. $Cu^{2+}(aq) + Mg(s) \rightarrow Mg^{2+}(aq) + Cu(s)$

2. $2\ Al(s) + 3\ Ni^{2+}(aq) \rightarrow 2\ Al^{3+}(aq) + 3\ Ni(s)$

3. no reaction

4. $2\ Li(s) + 2\ H_2O(l) \rightarrow 2\ Li^{+}(aq) + 2\ OH^{-}(aq) + H_2(g)$

5. $2\ H^{+}(aq) + Fe(s) \rightarrow Fe^{2+}(aq) + H_2(g)$

EXERCISE 4-7

1. Determine the moles of $Ca(OH)_2$, then use a mole ratio to determine moles of HNO_2, and finish by dividing the moles of HNO_2 by the liters of solution.

$$\left(\frac{0.1000\ \text{mol Ca(OH)}_2}{1\ L}\right)\left(\frac{10^{-3}\ L}{1\ mL}\right)(25.00\ mL)\left(\frac{2\ \text{mol HNO}_2}{1\ \text{mol Ca(OH)}_2}\right)\left(\frac{1}{30.00\ mL}\right)\left(\frac{1\ mL}{10^{-3}\ L}\right) =$$

0.1667 M HNO_2

2. Determine the moles of HNO_2, then use a mole ratio to determine moles of $Ca(OH)_2$, and finish by dividing the moles of $Ca(OH)_2$ by the liters of solution.

$$\left(\frac{0.1000\ \text{mol HNO}_2}{1\ L}\right)\left(\frac{10^{-3}\ L}{1\ mL}\right)(25.00\ mL)\left(\frac{1\ \text{mol Ca(OH)}_2}{2\ \text{mol HNO}_2}\right)\left(\frac{1}{30.00\ mL}\right)\left(\frac{1\ mL}{10^{-3}\ L}\right) =$$

0.04167 M $Ca(OH)_2$

3. Determine the moles of $Ca(OH)_2$, use a mole ratio to determine the moles of HNO_2, and finish with the molarity of the HNO_2 solution (inverted to get the correct units).

$$\left(\frac{0.01000 \text{ mol } Ca(OH)_2}{1 \text{ L}}\right)\left(\frac{10^{-3} \text{ L}}{1 \text{ mL}}\right)(45.00 \text{ mL})\left(\frac{2 \text{ mol } HNO_2}{1 \text{ mol } Ca(OH)_2}\right)\left(\frac{1 \text{ L}}{0.1000 \text{ mol } HNO_2}\right) =$$

9.000×10^{-3} L

4. Determine the moles of HNO_2, use a mole ratio to determine the moles of $Ca(OH)_2$, and finish with the molarity of the $Ca(OH)_2$ solution (inverted to get the correct units).

$$\left(\frac{0.2000 \text{ mol } HNO_2}{1 \text{ L}}\right)\left(\frac{10^{-3} \text{ L}}{1 \text{ mL}}\right)(40.00 \text{ mL})\left(\frac{1 \text{ mol } Ca(OH)_2}{2 \text{ mol } HNO_2}\right)\left(\frac{1 \text{ L}}{0.01500 \text{ mol } Ca(OH)_2}\right)$$

$= 0.2667$ L

5

Gases and Gas Laws

EXERCISE 5-1

1. $P_{Total} = P_A + P_B + P_C + \ldots$, where P_A and so on are the partial pressures of the separate gases in the mixture

2. $\dfrac{P_1 V_1}{T_1} = \dfrac{P_2 V_2}{T_2}$, where the subscripts refer to two sets of conditions

3. $PV = nRT$

4. 1 atm = 760 mm Hg = 760 torr = 1.01325×10^5 Pa (or 101.325 kPa)

5. $n = PV/RT$ If you did not get this answer, make sure you practice until you can.

6. $T_2 = \dfrac{(T_1)(P_2)(V_2)}{(P_1)(V_1)}$ If you did not get this answer, make sure you practice until you can.

7. If one variable changes, at least one other variable must change.
   a. **I** If the moles increase, there will be more molecules present to exert more pressure.
   b. **D** If the temperature decreases, the volume will decrease (Charles' law).
   c. **I** If the pressure increases, the temperature must also increase.
   d. **D** If the temperature decreases, the pressure must also decrease.
   e. **D** If the volume increases, the pressure must decrease (Boyle's law).

8. Rearrange the combined gas law with the pressure terms removed (no pressures given in the problem), and then enter the values (the temperature must be in kelvin units).

$$T_2 = \frac{(T_1)(V_2)}{(V_1)} = \frac{(300.\ K)(10.00\ L)}{(15.55\ L)} = 193\ K$$

Convert the Kelvin temperature to °C: 193 K − 273 = −80°C

9. Rearrange the combined gas law, and then enter the values (the temperature must be in Kelvin units); finally, add a volume and a pressure conversion.

$$P_2 = \frac{(P_1)(V_1)(T_2)}{(V_2)(T_1)} = \frac{(795.0\ torr)(1250.0\ mL)(298\ K)}{(1.000\ L)(273\ K)}\left(\frac{10^{-3}\ L}{1\ mL}\right)\left(\frac{1\ atm}{760\ torr}\right) =$$
1.43 atm

10. Rearrange the combined gas law, and then enter the values (the temperature must be in Kelvin units); finally, add a pressure conversion.

$$V_2 = \frac{(P_1)(V_1)(T_2)}{(P_2)(T_1)} = \frac{(225.0\ mmHg)(5.000\ L)(273\ K)}{(1.000\ atm)(300.\ K)}\left(\frac{1\ atm}{760\ mmHg}\right) =$$
1.35 L

11. Use Graham's law, one of the gases being O_2 and the other being unknown (it does not matter which if you are consistent). (Do not forget that oxygen is diatomic, which makes its molar mass = 32.00 g/mol.)

$$\frac{r_{O_2}}{r_{unk}} = \frac{0.3238 \text{ mL/s}}{0.1516 \text{ mL/s}} = \sqrt{\frac{M_{unk}}{32.00 \dfrac{g}{mol}}}$$

$M_{unk} = 146.0$ g/mol

12. Begin by rearranging the ideal gas equation to solve for moles ($= PV/RT$). Then enter the appropriate values (do not forget, you need the partial pressure of N_2, not the total pressure). (The partial pressure of nitrogen comes from Dalton's law, where the total pressure = partial pressure of N_2 + partial pressure of H_2O.) After using a pressure conversion, you have moles of N_2. Finish with a mole ratio from the balanced chemical equation and the molar mass of NaN_3.

$$\frac{\left[(875-24)\text{ torr}\right](8.25 \text{ L})}{\left(0.0821 \dfrac{L \cdot atm}{mol \cdot K}\right)(298 \text{ K})}\left(\frac{1 \text{ atm}}{760 \text{ torr}}\right)\left(\frac{2 \text{ mol NaN}_3}{3 \text{ mol N}_2}\right)\left(\frac{65.0 \text{ g NaN}_3}{1 \text{ mol NaN}_3}\right) =$$

16.4 g NaN_3

Thermochemistry

EXERCISE 6-1

1. **False**   To confirm this, pour two glasses of room temperature water into a bowl. The volume of water in the bowl is greater than the individual glasses (extensive), but the temperature remains the same (intensive).

2. **False**   1 nutritional Calorie = 1,000 calories (normal)

3. **True**   A calorimeter is the best method to determine the heat of a reaction.

4. **True**   A negative enthalpy change is exothermic, while a positive enthalpy change is endothermic.

5. **True**   The water lost heat, making the enthalpy change negative = exothermic.

6. **$1 J = 1 kg·m^2/s^2$**

7. **J/g·K (or J/g·°C)**   Either may be used.

8. **−488 kJ**   This is a Hess law problem:

Double: $C(s) + O_2(g) \rightarrow CO_2(g)$ $\Delta H = -393.5$ kJ

Double: $H_2(g) + 1/2\ O_2(g) \rightarrow H_2O(l)$ $\Delta H = -285.8$ kJ

Reverse: $HC_2H_3O_2(l) + 2\ O_2(g) \rightarrow 2\ CO_2(g) + 2\ H_2O(l)$ $\Delta H = -871$ kJ

This gives:

$2\ C(s) + 2\ O_2(g) \rightarrow 2\ CO_2(g)$ $\Delta H = 2(-393.5$ kJ$)$

$2\ H_2(g) + O_2(g) \rightarrow 2\ H_2O(l)$ $\Delta H = 2(-285.8$ kJ$)$

$2\ CO_2(g) + 2\ H_2O(l) \rightarrow HC_2H_3O_2(l) + 2\ O_2(g)$ $\Delta H = +871$ kJ

Cancel species on opposite sides and add the remaining:

$2\ C(s) + \cancel{2\ O_2(g)} \rightarrow \cancel{2\ CO_2(g)}$ $\Delta H = 2(-393.5$ kJ$)$

$2\ H_2(g) + O_2(g) \rightarrow \cancel{2\ H_2O(l)}$ $\Delta H = 2(-285.8$ kJ$)$

$\cancel{2\ CO_2(g)} + \cancel{2\ H_2O(l)} \rightarrow HC_2H_3O_2(l) + \cancel{2\ O_2(g)}$ $\Delta H = +871$ kJ

Sum:

$2\ C(s) + 2\ H_2(g) + O_2(g) \rightarrow HC_2H_3O_2(l)$ $\Delta H = -488$ kJ

Since the sum is the equation sought, the enthalpy change must be the energy change sought.

9. **−779.4 kJ**   This is a Hess law problem; solve like question 8:

Double: $N_2(g) + 3\,O_2(g) + H_2(g)$ → $2\,HNO_3(aq)$ $\Delta H = -413.14\,kJ$

Reverse and double: $N_2O_5(g) + H_2O(g)$ → $2\,HNO_3(aq)$ $\Delta H = 218.4\,kJ$

Reverse $2\,H_2(g) + O_2(g)$ → $2\,H_2O(g)$ $\Delta H = -483.64\,kJ$

This gives:

$2\,N_2(g) + 6\,O_2(g) + 2\,H_2(g) \rightarrow 4\,HNO_3(aq)$ $\Delta H = 2(-413.14\,kJ)$

$4\,HNO_3(aq)$ → $2\,N_2O_5(g) + 2\,H_2O(g)$ $\Delta H = 2(-218.4\,kJ)$

$2\,H_2O(g)$ → $2\,H_2(g) + O_2(g)$ $\Delta H = +483.64\,kJ$

Cancel species on opposite sides and add the remaining:

$2\,N_2(g) + \cancel{6}5\,O_2(g) + \cancel{2\,H_2(g)} \rightarrow \cancel{4\,HNO_3(aq)}$ $\Delta H = 2(-413.14\,kJ)$

$\cancel{4\,HNO_3(aq)}$ → $2\,N_2O_5(g) + \cancel{2\,H_2O(g)}$ $\Delta H = 2(-218.4\,kJ)$

$\cancel{2\,H_2O(g)}$ → $\cancel{2\,H_2(g)} + \cancel{O_2(g)}$ $\Delta H = +483.64\,kJ$

Sum:

$2\,N_2(g) + 5\,O_2(g) \rightarrow 2\,N_2O_5(g)$ $\Delta H = -779.4\,kJ$

Since the sum is the equation sought, the enthalpy change must be the energy change sought.

10. **−222.0 kJ/mol**   Convert the mass of C_3H_8 to moles using its molar mass. There are two components to the energy: the calorimeter and the water. The energy absorbed by the calorimeter is the heat capacity of the calorimeter times the temperature change, and the energy absorbed by the water is the mass of water times the specific heat of water (4.18 J/ mol K) times the temperature change. Sum the two energies and divide by the moles. The process (burning) is exothermic; therefore, make the final answer negative.

Electrons and Quantum Theory

EXERCISE 7-1

1. $E = h\nu$

2. An atom may only have one ground state.

3. The maximum numbers are $s = 2$, $p = 6$, $d = 10$, and $f = 14$.

4. The four quantum numbers and their symbols are n = principal quantum number, l = angular momentum quantum number, m_l = magnetic quantum number, and m_s = spin quantum number.

5. The sublevels within an orbital will half fill before electrons pair up in a sublevel.

6. a. **F** $1s^2 2s^2 2p^5$
    b. **Cu** $1s^2 2s^2 2p^6 3s^2 3p^6 4s^1 3d^{10}$ (this is an exception)
    c. **Re** $1s^2 2s^2 2p^6 3s^2 3p^6 4s^2 3d^{10} 4p^6 5s^2 4d^{10} 5p^6 6s^2 4f^{14} 5d^5$

7. Begin by rearranging the equation $c = \lambda\nu$:

$$\nu = \frac{c}{\lambda} = \left(\frac{3.00 \times 10^8 \text{ m / s}}{645 \text{ nm}}\right)\left(\frac{1 \text{ nm}}{10^{-9} \text{ m}}\right) = 4.65 \times 10^{14} \text{ s}^{-1}$$

8. Use the de Broglie relationship:

$$\lambda = \frac{h}{mv} = \left[\frac{(6.63 \times 10^{-34} \text{ J} \cdot \text{s})}{(6.6 \times 10^{-24} \text{ g})\left(5.5 \times 10^{10} \frac{\text{m}}{\text{h}}\right)}\right]\left(\frac{\frac{\text{kg} \cdot \text{m}^2}{\text{s}^2}}{\text{J}}\right)\left(\frac{3600 \text{ s}}{1 \text{ h}}\right)\left(\frac{10^3 \text{ g}}{1 \text{ kg}}\right)$$

$$= 6.6 \times 10^{-15} \text{ m}$$

The values within the brackets are the values for the de Broglie equation. The values outside the brackets are conversion factors.

9. Begin by rearranging the equation:

$$\lambda = \frac{hc}{E} = \left| \frac{\left(6.63 \times 10^{-34} \text{ J} \cdot \text{s}\right)\left(3.00 \times 10^8 \frac{\text{m}}{\text{s}}\right)}{\left(239 \frac{\text{kJ}}{\text{mol}}\right)}\right| \left(\frac{1 \text{ kJ}}{10^3 \text{ J}}\right)\left(\frac{6.022 \times 10^{23}}{\text{mol}}\right)$$

$$= 5.01 \times 10^{-7} \text{ m}$$

The values within the brackets are the values for the rearranged equation. The values outside the brackets are a conversion factor and Avogadro's number.

8

Periodic Trends

EXERCISE 8-1

1. The effective nuclear charge is the nuclear charge minus the screening effect of the core electrons.

2. Columns on the periodic table are groups or families.

3. The effective nuclear charge increases toward the right in any period of the periodic table.

4. **False** Energy is always needed (endothermic) to remove an electron (ionize).

5. **True** These are especially stable arrangements; therefore, additional energy is necessary.

6. **c** Removing an electron from a gaseous atom produces a cation (positive ion).

7. The increase in the effective nuclear charge toward the right of a period causes the ionization energy to increase toward the right.

8. The atomic radius increases toward the bottom of any family. This increase in radius leads to a decrease in the ionization energy toward the bottom of the family.

9. **Mg, P, and Ar**   These are the elements with filled or half-filled shells or subshells.

10. K^+ is smaller than K (K^+ has fewer electrons [lost its $4s$ subshell] and the positive charge pulls the remaining electrons closer to the nucleus.)

11. **True**   The first electron affinity may be either.

12. **a**   Adding an electron to a gaseous atom produces an anion (negative ion).

13. The increase in the effective nuclear charge toward the right of a period causes the electron affinities to increase toward the right, except for the noble gases.

14. The atomic radius increases toward the bottom of any family; this increase in radius leads to a decrease in the electron affinity toward the bottom of the family.

15. **Cl^- is larger than Cl.**   There is greater repulsion due to the increased number of electrons.

16. **K < Br < Ar**   In general, the ionization energies increase toward the right in a period.

17. **Ar < K < Br**   Noble gases (Ar) have no electron affinity; for the other two elements, electron affinity increases toward the right in a period.

18. **Mg^{2+} < Na^+ < Ne < F^- < O^{2-}**   All these species are isoelectronic; therefore, if there were other factors, they would all be the same size. However, there is another factor, the charge. The greater the positive charge, the smaller the result. The greater the negative charge, the larger the result.

19. The energy necessary will be the sum of the first, second, and third ionization energies.

20. It is an exception because moving an electron to the $3d$ subshell changes the configuration from a less stable $3d^4$ to a more stable half-filled $3d^5$.

Lewis Structures and Chemical Bonding

EXERCISE 9-1

1. During chemical reactions, atoms tend to gain, lose, or share electrons to achieve an octet of electrons in their outer shell.

2. *Isoelectronic* means that the species have the same number and arrangement of electrons.

3. The number may be determined from the electron configuration, with the number of outer shell electrons being the ones indicated. However, for the representative elements, such as these, it is faster to locate the element on the periodic table and use the old numbering system (A designation), as the number preceding the A is the number of valence electrons.

   a. **1** Column 1A
   b. **8** Column 8A
   c. **7** Column 7A
   d. **5** Column 5A
   e. **4** Column 4A

4. Electronegativity is a measure of the attractive force that an atom in a compound exerts on electrons in a bond.

5. **e** Of the pairs listed, these two elements are the farthest apart on the periodic table.

6. **b** O^- needs one more electron to achieve its octet. If you are unsure about any of the others, write the electron configuration of the atoms and remove/add the number of electrons indicated by the charge.

7.   a. **2** S has 6 valence electrons, so 2 more electrons are required.
      b. **1** I has 7 valence electrons, so 1 more electron is required.
      c. **0** Ar is a noble gas (other than He), so it already has an octet.

8. a. **1** Cs has 1 valence electron, so losing this electron will make the ion isoelectronic with a noble gas (octet).

b. **2** Mg has 2 valence electrons, so losing 2 electrons will make the ion isoelectronic with a noble gas (octet).

c. **7** Cl has 7 valence electrons, so losing 7 electrons will make the ion isoelectronic with a noble gas (octet).

9. The lattice energy is defined as the energy required to separate the ions in one mole of an ionic solid.

10. **d** The representative elements will never have more than three bonds between two atoms.

11. **a** As long as there is an electronegativity difference, there will be some polarity.

12. **b** H can only form one bond, so it cannot connect to more than one other atom.

13. **N-N > N=N > N≡N** The more bonds present between two atoms, the shorter the bond. There must be identical atoms involved in each comparison, meaning that a comparison of N=N to N=O cannot be done by inspection, but comparing N-O to N=O is allowed.

14. **C-O < C=O < C≡O** In general, the more bonds, the more energy required to break the bonds.

15. The problem says to react atoms (not molecules), so begin with the individual atoms. A reaction requires reactants, a reaction arrow, and products. A metal reacts with a nonmetal to produce ions. These are the Lewis symbols for the formation of LiF.

$$Li\cdot \quad + \quad :\overset{\cdot\cdot}{\underset{\cdot\cdot}{F}}\cdot \quad \longrightarrow \quad Li^+ \quad + \quad \left[:\overset{\cdot\cdot}{\underset{\cdot\cdot}{F}}:\right]^-$$

16. The problem says to react atoms (not molecules), so begin with the individual atoms. A reaction requires reactants, a reaction arrow, and products. A metal reacts with a nonmetal to produce ions. These are the Lewis symbols for the formation of CaF_2.

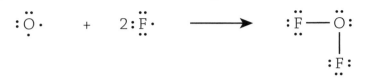

17. The problem says to react atoms (not molecules), so begin with the individual atoms. A reaction requires reactants, a reaction arrow, and products. Two nonmetals combine to produce a molecule. These are the Lewis symbols for the formation of OF_2.

$$:\ddot{O}\cdot \quad + \quad 2:\ddot{F}\cdot \quad \longrightarrow \quad :\ddot{F}-\ddot{O}: \\ | \\ :\ddot{F}:$$

18. The oxygen needs more electrons to complete its octet; therefore, it is the central atom (also since this is an oxyacid, it is necessary to attach the acid H to the O and the O to the other element present). This is the Lewis structure of HOCl.

$$H-\ddot{O}: \\ | \\ :\ddot{Cl}:$$

19. Sodium phosphate is an ionic compound; therefore, the Lewis structure must show ions. The ions are Na^+ and PO_4^{3-}. This is the Lewis structure of Na_3PO_4.

$$3\ Na^+ \quad \left[:\ddot{O}-\overset{\overset{\displaystyle :\ddot{O}:}{|}}{\underset{\underset{\displaystyle :\ddot{O}:}{|}}{P}}-\ddot{O}: \right]^{3-}$$

EXERCISE 9-2

All atoms except H are obeying the octet rule.

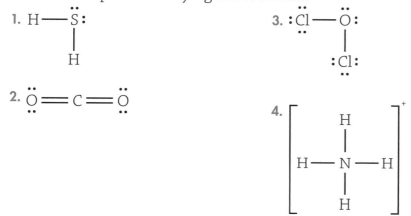

EXERCISE 9-3

In all cases, other than *d,* the central atom is not obeying the octet rule. All noncentral atoms are obeying the octet rule.

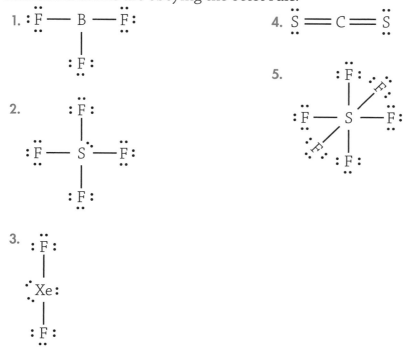

EXERCISE 9-4

The resonance structure may be drawn in any order. NO_2 has an odd number of electrons, so the less electronegative atom (N) has less than an octet (all other atoms in all structures have an octet).

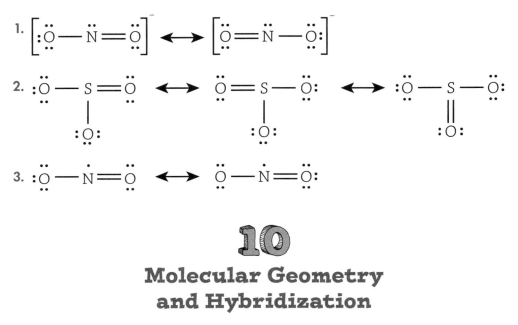

10
Molecular Geometry and Hybridization

EXERCISE 10-1

1. Valence-shell electron pair repulsion

2. See the figure in the chapter. The better you can reproduce this table, the better you understand the material and the fewer times you will need to look back at the table (saving time).

   See the problems at the end of Chapter 9 for the Lewis structures needed for questions 3–11. (You may want to redraw the Lewis structures here for practice.) In each case, determine the total number of electron pairs (multiple bonds count as 1 pair) around the central atom and the number of those pairs that are lone pairs. Then use the figures in the text, either from memory (best) or look back at them.

3. **a.** bent   **b.** bent   **c.** trigonal planar

4. **a.** distorted tetrahedral   **b.** linear   **c.** tetrahedral

5. **a.** octahedral   **b.** linear

6. **a.** sp^3 **b.** sp^3 **c.** sp^2

7. **a.** sp^3d **b.** sp^3d **c.** sp^3

8. **a.** sp^3d^2 **b.** sp

9. **a.** polar   **b.** polar   **c.** nonpolar

10. **a.** polar   **b.** nonpolar

11. **a.** nonpolar   **b.** nonpolar

11
Solids, Liquids, and Intermolecular Forces

EXERCISE 11-1

1. **c**   Burning is a chemical reaction.

2. **d**   Ion-dipole forces require two substances, in this case, one of the substances (NaCl) produces ions and the other H_2O (gives the dipole).

3. **b**   This is an ionic compound, not a polar molecule.

4. **a**   There is no H attached to an N, O, or F. All Hs are attached to the C.

5. **e**   This is the only nonpolar substance; therefore, this is the only one in the group with no other intermolecular forces besides London dispersion forces.

6. **surface tension**   This produces the "skin" on the surface of a liquid.

7. resistance of a substance to flow

8. spontaneous rising of a liquid through a narrow tube against the force of gravity

9. **hydrogen bonding**   A strong intermolecular force to overcome.

10. extensive ordering of the particles

11. a very regular ordering of the particles in a three-dimensional structure called the crystal lattice

12. **b**   Because it contains H_2O molecules.

13. **e**   Because every carbon atom is covalently bonded to four carbon atoms to form a network.

14. **a**   Because only Xe atoms are present.

15. **c**   Because there are sodium ions and chloride ions.

16. **d**   Because sodium is a metal.

17. See the phase diagram in Chapter 1. As a sketch, your diagram does not need to be perfect.

18. triple point and the critical point

19.   a. **dipole-dipole**   because this is a polar molecule
   b. **hydrogen bonding**   because there is H attached to N
   c. **covalent bonding**   because the C atoms are connected to each other by covalent bonds
   d. **ionic bonding**   because there are ions present
   e. **London dispersion force**   because the molecules are nonpolar

20.   a. **dipole-dipole**   because the molecules are polar
   b. **hydrogen bonding**   because there is H attached to O
   c. **dipole-dipole**   because the molecules are polar (no hydrogen bonding because H is attached to C not F)
   d. **ionic bonding**   because this is an ionic compound
   e. **London dispersion force**   because this is a nonpolar molecule

12
Solutions

1. "Like dissolves like"

2. **d**   A saturated solution contains the maximum amount of solute that will normally dissolve.

3. **c**   Molality only has the solvent in the denominator.

4. **a**   Dilution involves changing the quantity of solute, which changes the volume, density, osmotic pressure, and many other factors. No change in the quantity of solute occurs.

5. **b**   Raoult's law uses mole fractions, osmotic pressure uses molarity, and freezing point depression uses molality.

6. **a**   The Tyndall effect is the scattering of light by a colloid. Only colloids exhibit the Tyndall effect, all combination can display the other properties (in principle).

7. The mass percent will use the mass of a substance (numerator), mass of the solution (denominator) and 100%.

$$\text{mass \%} = \frac{(10.0 \text{ g table salt})}{(10.0 + 15.0 + 500.0) \text{ g solution}} \times 100\% = 1.90\%$$

8. The molarity needs the moles of solute (from the mass and the molar mass) and the kilograms of the solvent (mass and a conversion).

$$\text{molality} = \left(\frac{50.0 \text{ g C}_2\text{H}_5\text{OH}}{250.0 \text{ g H}_2\text{O}}\right)\left(\frac{1000 \text{ g}}{1 \text{ kg}}\right)\left(\frac{1 \text{ mol C}_2\text{H}_5\text{OH}}{46.1 \text{ C}_2\text{H}_5\text{OH}}\right) =$$
$$4.34 \text{ m C}_2\text{H}_5\text{OH}$$

9. This problem involves Raoult's law: $P_{\text{solution}} = X_{\text{solvent}} P^\circ_{\text{solvent}} + X_{\text{solute}} P^\circ_{\text{solute}}$

$P_{\text{solution}} = ?, X_{\text{solvent}} = 1 - X_{\text{solute}}, P^\circ_{\text{solvent}} = 114 \text{ torr}, X_{\text{solute}} = 0.250, P^\circ_{\text{solute}}$
$= 197 \text{ torr}$

$P_{\text{solution}} = (1.000 - 0.250)(114 \text{ torr}) + (0.250)(197 \text{ torr}) = 135 \text{ torr}$

10. This is a colligative property problem involving an electrolyte; therefore, it is necessary to determine the van't Hoff factor for $(NH_4)_3PO_4$ (i = **3** NH_4^+ + **1** PO_4^{3-} = 4). Enter the van't Hoff factor, the given constants, and the molality into the freezing point depression or boiling point elevation equation to find ΔT. Subtract ΔT from the freezing point depression equation from the normal freezing point. Add ΔT from the boiling point elevation equation to the normal boiling point.

$$\Delta T_f = iK_f m = (4)\,(1.86°C/\text{molal})\,(1.50\text{ m}) = 11.2°C;\ T_f = (0.00 - 11.2)°C$$
$$= -11.2°C$$

$$\Delta T_b = iK_b m = (4)\,(0.52°C/\text{molal})\,(1.50\text{ m}) = 3.12°C;\ T_b = (100.00 + 3.12)$$
$$= 103.12°C$$

11. Rearrange the osmotic pressure equation ($\pi V = nRT$) to find moles, enter the given values, and a pressure conversion. The molar mass is the given grams divided by the moles calculated.

$$\left[\frac{(0.195\text{ torr})\,(0.0300\text{ L})}{\left(0.0821\ \dfrac{\text{L}\cdot\text{atm}}{\text{mol}\cdot\text{K}}\right)(298.2\text{ K})}\right]\left(\frac{1\text{ atm}}{760\text{ torr}}\right) = 3.14\times10^{-7}\text{ mol};$$

$$\frac{\left(6.95\times10^{-3}\text{ g}\right)}{\left(3.14\times10^{-7}\text{ mol}\right)} = 2.21\times10^4\text{ g/mol}$$

13
Kinetics

EXERCISE 13-1

1. The five factors are (1) nature of the reactants, (2) the temperature, (3) the concentration of the reactants, (4) physical state of the reactants, and (5) catalysts.

2. 2 ($9 = 3^2$)

3. 0 ($1 = 9^0$)

4. $t_{1/2} = 1/k\,[A]_0$ (Make sure you use the correct half-life equation.)

5. Activation energies have large positive values; this answer is neither large nor positive.

6. True   A heterogeneous catalyst is in a different phase than the reactants. An automobile catalytic converter is a solid and the reactants are gases.

7. False   Rate laws come from experimentation.

8. True   The rate-determining step in a mechanism can yield a rate law (experiments are necessary to determine the mechanism).

9. False   The probability of three molecules colliding simultaneously with the necessary amount of energy is very unlikely. (So unlikely as to make it impossible.)

10. False   If the step involved 2 A and 1 B colliding, it would be a termolecular step, which is very unlikely to occur.

11. This is a first-order reaction, so begin with the first-order integrated rate law. Rearrange the integrated rate law and enter the given values:

$$t = \left(\frac{\ln\dfrac{[1.0\text{ M}]}{[0.38\text{ M}]}}{1.25\text{ hr}^{-1}}\right) = 0.77\text{ hr}$$

12. This is a second-order reaction, so begin with the second-order half-life equation. Rearrange the half-life equation and enter the given values.

$$t_{1/2} = \frac{1}{(1.78\text{ M}^{-1}\text{ s}^{-1})[0.575\text{ M}]} = 0.977\text{ s}$$

13. Assign the two k and T values into pairs with matching subscripts (in the following, $T_1 = 100°C = 373$ K, if you chose to make this value T_1, and correctly match the other subscripts, you will still get the same answer). Make sure the temperatures are in K. Enter the values into the modified Arrhenius equation and solve.

$$\ln\frac{9.3\times10^{-5}\text{ M}^{-1}\text{s}^{-1}}{1.0\times10^{-3}\text{ M}^{-1}\text{s}^{-1}} = \frac{E_a}{8.314\text{ J/mol}\cdot\text{K}}\left[\frac{1}{403\text{ K}} - \frac{1}{373\text{ K}}\right]\left(\frac{1\text{ kJ}}{1000\text{ J}}\right)$$

$E_a = 99$ kJ/mole

When using this equation, any rounding may lead to serious errors. Many people make an error when calculating the ln value; the unrounded value should be -2.375155786 (no units), but if you reverse the ks, the sign will be positive. The other place where many people make an error is in the T calculation in the brackets, the unrounded value should be $-1.995755693 \times 10^{-4}$ K^{-1}, but if you reversed the Ts, this value will be positive.

14
Chemical Equilibria

EXERCISE 14-1

1. **pressure or concentration**   but not a mixture

2. **products**   The arrangement is always products (numerator) over reactants (denominator).

3. **left**   The same as for any chemical equation.

4. **False**   Catalysts increase the rates of reactions; they do not alter the position of the equilibria involved.

5. **False**   There are equal numbers of moles of gas on each side of the equilibrium arrow, so changes in volume/pressure have no effect.

6. **True**   Changing the amount of a solid has no effect (as long as some solid is present).

7. **True**   Equilibrium constant expressions do not include solids, liquids, or solvents.

8. **False**   Water is the solvent; solvents do not appear in equilibrium constant expressions.

9. Enter the appropriate values into the relationship relating K_p and K_c. While the units are listed in the following calculation, they will be neglected. The exponent comes from the three moles of gas being formed from no moles of gas.

$$K_p = 1.58 \times 10^{-8} \left[\left(0.0821 \, \frac{L \cdot atm}{mol \cdot K} \right) (523 \text{ K}) \right]^{(3-0)} = 1.25 \times 10^{-3}$$

10. Head each column with the identity of each substance (it will help to use the same order as in the equilibrium chemical equation). Enter the given concentrations into the "Initial" row below each substance, with a 0 for anything not given. The equilibrium will shift toward the side with one or more 0 values. (If there are no 0 values, it is possible to determine the direction by entering the values into a reaction quotient [Q] calculation.) In this case, the reaction goes to the right. The "Change" line includes x values, with + values on the side the equilibrium is shifting toward and − values on the opposite side. Every x is preceded by a number equal to the substance's coefficient in the reaction. The "Equilibrium" line is the sum of the information in the first two lines.

	NO	**H$_2$**	**N$_2$**	**H$_2$O**
initial	0.250 M	0.100 M	0	0.200 M
change	− 2x	− 2x	+ x	+ 2x
equilibrium	0.250 − 2x	0.100 − 2x	+ x	0.200 + 2x

11. The procedure here is identical to that in question 12, with pressures being used instead of concentrations. Leave out the solid (C).

	S_2	C	CS_2
initial	0.431 atm	—	0
change	$-x$	—	$+x$
equilibrium	$0.431 - x$	—	$+x$

12. No matter what, do not stop working this problem before writing the equilibrium constant expression. Then construct an ICE table (the equilibrium amount of $C_2H_2 = x = 0.01375$ M). Substitute 0.01375 for all x terms and then enter the values into the equilibrium constant expression and calculate the answer.

$$K_c = \frac{[C_2H_2][H_2]^3}{[CH_4]^2} = \frac{[0.01375][3\,(0.01375)]^3}{[0.0300 - 2\,(0.01375)]^2} = 0.154$$

13. No matter what, do not stop working this problem before writing the equilibrium constant expression (leave out the solid). Then construct an ICE table, enter the x values, and solve for the partial pressures.

$$K_p = P_{PH_3}\,P_{BCl_3} = (x)\,(x) = 1.60\,P_{PH_3} = P_{BCl_3} = 1.26 \text{ atm}$$

14. No matter what, do not stop working this problem before writing the equilibrium constant expression. Construct an ICE table and enter the equilibrium line into the equilibrium constant expression. Assume $0.200 - 2x \approx 0.200$ to simplify the calculation and solve for x and enter the values into the equilibrium line of the ICE table.

$$K_c = \frac{[H_2]^2\,[S_2]}{[H_2S]^2} = 9.1 \times 10^{-8}$$

$[H_2] = 2x = 1.9 \times 10^{-3}$ M, $[S_2] = x = 9.7 \times 10^{-4}$ M, and $[H_2S] = 0.200 - 2x = 0.198$ M

The x is sufficiently small to justify the assumption.

EXERCISE 14-2

The expressions are the concentrations of the products over the concentrations of the reactants with the coefficients being exponents. Do not include solids, liquids, and solvents.

1. $K_c = \dfrac{[CO_2][H_2O]^2}{[CH_4][O_2]^2}$

2. $K_c = \dfrac{[CO_2]^3}{[CO]^3}$

3. $K_c = \dfrac{[H^+][NO_2^-]}{[HNO_2]}$

4. $K_c = \dfrac{\left[Cu(NH_3)_4^{2+}\right]}{[Cu^{2+}][NH_3]^4}$

5. $K_c = [Zn^{2+}][CO_3^{2-}]$

EXERCISE 14-3

The expressions are the partial pressures of the products over the partial pressures of the reactants with the coefficients being exponents. Include only gases.

1. $K_p = \dfrac{P_{N_2}\, P_{O_2}^2}{P_{NO_2}^2}$

2. $K_p = \dfrac{P_{CO}\, P_{H_2}}{P_{H_2O}}$

3. $K_p = \dfrac{1}{P_{O_2}}$

4. $K_p = \dfrac{P_{CO_2}^8}{P_{C_4H_{10}}^2\, P_{O_2}^{13}}$

5. $K_p = P_{CO_2}$

EXERCISE 14-4

This exercise involves an application of Le Châtelier's principle, so no calculations are necessary. The focus, in this case, is the amount of hydrogen, not the direction.

1. **decrease** Adding a reactant will cause a shift to the right using up the reactants, including the hydrogen.

2. **no change** The water is a liquid; liquids do not alter equilibria (as long as some is present).

3. **no change** A catalyst alters the rate of the reaction, not the position of the equilibrium.

4. **decrease** This is an endothermic reaction (ΔH = +); therefore, adding heat will shift the process to the right using hydrogen (and CO_2).

5. **decrease** Increasing the pressure will shift the equilibrium toward the side with less moles of gas (right, in this case). Shifting to the right will decrease the amount of hydrogen.

15

Acids and Bases

EXERCISE 15-1

1. chloric acid, $HClO_3$; hydrobromic acid, HBr; hydrochloric acid, HCl; hydroiodic acid, HI; nitric acid, HNO_3; perchloric acid, $HClO_4$; sulfuric acid, H_2SO_4

2. alkali metal (Group IA) hydroxides (LiOH, NaOH, KOH, RbOH, CsOH); calcium hydroxide, $Ca(OH)_2$; strontium hydroxide, $Sr(OH)_2$; and barium hydroxide, $Ba(OH)_2$

3. $K_w = K_a K_b = 1.0 \times 10^{-14}$ (This applies to any conjugate acid-base pair.)

4. $pH = -\log [H^+]$

5. $pK_w = pH + pOH = 14.00$

6. The equilibrium is:

$$NH_3(aq) + H_2O(l) \rightleftharpoons NH_4^+(aq) + OH^-(aq) \quad K_b = 1.8 \times 10^{-5}$$

Use a generic K_b expression, construct an ICE table, and solve for $x = [OH^-]$. Once $[OH^-]$ is known, use one or more of the following: $pOH = -\log [OH^-]$, $pK_w = pH + pOH = 14.000$, $pH = -\log [H^+]$, or $K_w = [H^+] [OH^-] = 1.00 \times 10^{-14}$.

$$[OH^-] = 1.3 \times 10^{-3} \text{ M, pOH} = 2.87, pH = 11.13$$

EXERCISE 15-2

Remove 1 H^+ from each formula given (do not forget to adjust the charge on the conjugate base to reflect the removal of 1 H^+).

1. NO_3^-

2. $C_2H_3O_2^-$

3. NH_2^-

4. HSO_4^-

5. CO_3^{2-}

EXERCISE 15-3

Add 1 H^+ to each formula given (do not forget to adjust the charge on the conjugate acid to reflect the addition of 1 H^+).

1. HCl

2. HPO_4^{2-}

3. NH_4^+

4. HNO_2

5. OH^-

EXERCISE 15-4

1. $pH = -\log[H^+] = 3$
2. $pH = -\log[H^+] = 4.82$
3. $pH = -\log[H^+] = 0.00$
4. $pH = -\log[H^+] = 15.00$
5. $pH = 9.8$ (Use $pK_w = pH + pOH = 14.000$)

EXERCISE 15-5

Use one or more of the following: $[H^+] = 10^{-pH}$, $pK_w = pH + pOH = 14.000$, $[OH^-] = 10^{-pOH}$, or $K_w = [H^+][OH^-] = 1.00 \times 10^{-14}$ (For more practice, try determining each value by two different methods.)

1. $[H^+] = 10^{-4}$ M
2. $[H^+] = 10^{-3.75} = 1.8 \times 10^{-4}$ M
3. $[H^+] = 1.0 \times 10^{-7}$ M
4. $[H^+] = 3.1 \times 10^{-12}$ M
5. $[H^+] = 3.2$ M

EXERCISE 15-6

These are all strong acids or bases, so it is possible to go directly from the concentrations given to $[H^+]$ or $[OH^-]$.

1. $pH = -\log[H^+] = -\log[1.0 \times 10^{-2}] = 2.00$
2. $pH = -\log[H^+] = -\log[4.5 = 10^{-5}] = 4.35$
3. $pOH = -\log[OH^-] = -\log[1.5 \times 10^{-3}] = 2.82$; $pH = 14.000 - pOH = 11.18$

EXERCISE 15-7

All of these require the relationship: $K_a = \dfrac{[H^+][A^-]}{[HA]}$. Use the K_a given and create an ICE table. It is possible to simplify questions 1 and 3 by neglecting the $-x$ and avoid the quadratic.

1. $[H^+] = 1.3 \times 10^{-3}$ M
2. $[H^+] = 1.7 \times 10^{-2}$ M (quadratic)
3. $[H^+] = 8.5 \times 10^{-6}$ M

EXERCISE 15-8

Enter the answers from 15-7 into: $pH = -\log [H^+]$.

1. $pH = 2.87$
2. $pH = 1.76$
3. $pH = 5.07$

16
Buffers and Additional Equilibria

EXERCISE 16-1

1. These are the two different forms of the Henderson–Hasselbalch equation:

$$pH = pK_a + \log\frac{[A^-]}{[HA]} \qquad pOH = pK_b + \log\frac{[HA]}{[A^-]}$$

2. If the HA is given, remove 1 H^+ to get the A^-. If the A^- is given, add 1 H^+ to get the HA.

[HA]	[A⁻]
HNO_2	NO_2^-
NH_4^+	NH_3
HCO_3^-	CO_3^{2-}
$H_2PO_4^-$	HPO_4^{2-}
$CH_3NH_3^+$	CH_3NH_2

3. As a K_f the *only* way to write the equilibrium is $Ag^+(aq) + 2\ NH_3(aq) \rightleftarrows$ $[Ag(NH_3)_2]^+(aq)$. From this equilibrium, write the expression as products over reactants with the coefficients being exponents.

$$K_f = \frac{\left[Ag(NH_3)_2^+\right]}{\left[Ag^+\right]\left[NH_3\right]^2}$$

EXERCISE 16-2

These are heterogeneous equilibria, so leave out the solid (denominator). The answers only contain the ions from the solid raised to a power equal to the coefficient in the equilibrium.

1. $K_{sp} = [Ag^+]\,[Br^-]$
2. $K_{sp} = [Ag^+]^2[SO_4^{2-}]$
3. $K_{sp} = [Al^{3+}]\,[F^-]^3$
4. $K_{sp} = [Ca^{2+}]\,[CO_3^{2-}]$
5. $K_{sp} = [Ca^{2+}]^3[PO_4^{3-}]^2$

EXERCISE 16-3

The equilibrium is $CaCO_3(s) \rightleftarrows Ca^{2+}(aq) + CO_3^{2-}(aq)$.

$$K_{sp} = [Ca^{2+}]\,[CO_3^{2-}] = 8.7 \times 10^{-9}$$

1. $K_{sp} = [x]\,[x] = 8.7 \times 10^{-9}$ and solve for $x = 9.3 \times 10^{-5}\ M$.

2. Ca^{2+} is a spectator ion:

$K_{sp} = [6.5 \times 10^{-2}\ M\ Ca^{2+} + x]\,[x] = 8.7 \times 10^{-9}$ and solve for $x = 1.4 \times 10^{-7}\ M$.

3. CO_3^{2-} is a spectator ion:

$K_{sp} = [x]\,[0.35 + x] = 8.7 \times 10^{-9}$ and solve for $x = 2.5 \times 10^{-8}\ M$.

EXERCISE 16-4

These are buffer solutions, so use the Henderson–Hasselbalch equation. You may need to use one or more of the following: $pK_a = -\log K_a$, $pK_b = -\log K_b$, $pH + pOH = pK_w = 14.000 = pK_a + pK_b$, $K_w = [H^+][OH^-] = 1.00 \times 10^{-14}$.

1. $pH = 4.76$

2. $pH = 9.16$

3. $pH = 3.04$

4. $pH = 5.38$

5. $pH = 10.44$ (only pK_{a2} is necessary)

   Exercises 7 and 8 involve titration problems. The reactions are:

   $$Sr(OH)_2(aq) + 2\ HCl(aq) \rightarrow Sr^{2+}(aq) + 2\ Cl^-(aq) + 2\ H_2O(l)$$
   $$NH_3(aq) + HCl(aq) \rightarrow NH_4^+(aq) + Cl^-(aq)$$

   In each case, calculate the original moles and the moles added in each step. In all parts, except (a), a limiting reactant determination is necessary. From the limiting reactant, it is possible to determine the amount of each species after the addition. The next step will depend on what remains after the reaction. When both reactants are limiting, it is the equivalence point. Do not forget, as solution is added, the volume changes, which causes the concentrations to change.

EXERCISE 16-5

This exercise involves a strong acid–strong base titration; therefore, there are no equilibrium calculations. The amount of acid or base remaining goes into either $pOH = -\log [OH^-]$ or $pH = -\log [H^+]$, and sometimes $pH + pOH = 14.000 = pK_w$.

1. $pH = 13.079$

2. $pH = 12.602$

3. $pH = 7.000$

4. $pH = 1.620$

EXERCISE 16-6

This exercise involves a strong acid–weak base titration; therefore, anytime NH_3 remains, there will be a K_b calculation, and when NH_4^+ is present, there will be a K_a calculation (unless the strong acid (HCl) is in excess). (a) is a simple K_b problem. The next calculation is a buffer ($pOH = pK_b + \log\dfrac{[HA]}{[A^-]}$).

The next calculation is at the equivalence point (a generic K_a calculation). The final calculation uses the excess strong acid ($pH = -\log[H^+]$).

1. pH = 11.14
2. pH = 9.25
3. pH = 5.25
4. pH = 1.66

17
Entropy and Free Energy

EXERCISE 17-1

1. The first law of thermodynamics states that the total energy of the universe is constant. The second law of thermodynamics states that all processes that occur spontaneously move in the direction of an increase in entropy of the universe (system + surroundings). The third law of thermodynamics states that for a pure crystalline substance at 0 K the entropy is zero.

2. **a.** Going from a solid to a liquid involves an increase in entropy as would a change from a liquid to a gas or a solid to a gas.

3. negative

4. $G = H - TS$

5. If $\Delta S_{\text{surroundings}}$ is sufficiently positive to make the total entropy change positive, the process will be spontaneous.

6. Standard conditions in this book are: 25°C (298 K), 1 atm, and 1 M.

7. $\Delta G = \Delta G° + RT \ln Q$

8. $\Delta G° = -RT \ln K$

9. $R = 8.314$ J/mol·K (This R is simplest to use because energy [joules] is needed in most cases.)

10. The setup of K and Q is the same. K uses equilibrium quantities, whereas Q uses nonequilibrium, often initial, quantities.

11. In both cases, the values are zero. (In the case, of a solid element A, the reaction would be A(s) → A(s).)

12. The calculation is products minus reactants, with the appropriate numbers (given) entered:

$\Delta S° = $ [(1 mol CaSO$_4$·2H$_2$O) (194.0 J/mol·K] – [(1 mol Ca) (41.63 J/mol·K) + (1 mol S) (31.88 J/mol·K) + (2 mol H$_2$) (131.0 J/mol·K) + (3 mol O$_2$) (205.0 J/mol·K)]

$\Delta S° = -757$ J/K

13. The calculation is products minus reactants, with the appropriate numbers (given) entered:

$\Delta G° = $ [(1 mol CaSO$_4$) (–1320.3 kJ/mol) + (2 mol H$_2$O) (–237.2 kJ/mol) + (1 mol SO$_2$) (–300.4 kJ/mol)] – [(1 mol Ca) (0.00 kJ/mol) + (2 mol H$_2$SO$_4$) (–689.9 kJ/mol)]

$\Delta G° = -715.3$ kJ

14. This is an equilibrium process so set $\Delta G = \Delta H - T\Delta S$ equal to 0, rearrange, and solve for T. (Do not forget that a conversion is necessary because ΔH is in kJ and ΔS is in J.)

$$T = \frac{\Delta H}{\Delta S} = \left(\frac{31.9 \text{ kJ/mol}}{96.8 \text{ J/mol} \cdot \text{K}}\right)\left(\frac{10^3 \text{ J}}{1 \text{ kJ}}\right) = 3.30 \times 10^2 \text{ K}$$

15. Enter the appropriate values into $\Delta G° = -RT \ln K$ and solve.

$$\Delta G° = -\left(\frac{8.314 \text{ J}}{\text{mol} \cdot \text{K}}\right)(298 \text{ K}) \ln (1.9 \times 10^{-16}) = 8.97 \times 10^4 \text{ J/mol}$$

16. Enter the appropriate values into $\Delta G = \Delta G° + RT \ln Q$, where Q is products/reactants.

$$\Delta G = 120.0 \text{ kJ} + (8.314 \text{ J/mol·K})(298 \text{ K})\left(\frac{1 \text{ kJ}}{10^3 \text{ J}}\right) \ln \frac{[2.50]^2}{[1.50]} = 123.5 \text{ kJ}$$

Do not forget to ignore the solid.

18
Electrochemistry

EXERCISE 18-1

1. Oxidation is the loss of electrons and reduction is the gain of electrons.

2. These are all compounds; therefore, the sum of the oxidation numbers MUST be 0.
    a. $H = +1$ and $O = -2$
    b. $Na = +1$, $S = +6$, and $O = -2$
    c. $Fe = +3$ and $Cl = -1$
    d. $K = +1$ and $O = -1$
    e. $Ca = +2$, $Cr = +6$, and $O = -2$

3. **a.** negative   **b.** positive

4. **a.** cathode   **b.** cathode (Reduction is always at the cathode.)

5. The oxidizing agent is reduced.

6. A faraday is a mole of electrons.

7. **a.** $2 Cl^-$ **b.** Cr^{3+} **c.** $3 e^-$ **d.** $2 e^-$ **e.** $2 e^-$

8. This is a stoichiometry conversion problem based on the half-reaction:
$Mg^{2+} + 2\ e^- \rightarrow Mg$.

grams Mg =

$$(550\ \text{min})\left(\frac{60\ \text{s}}{1\ \text{min}}\right)\left(\frac{3.50\ \text{coul}}{\text{s}}\right)\left(\frac{1\ \text{mol}\ e^-}{96485\ \text{coul}}\right)\left(\frac{1\ \text{mol Mg}}{2\ \text{mol}\ e^-}\right)\left(\frac{24.31\ \text{g Mg}}{1\ \text{mol Mg}}\right)$$

= 14.6 g Mg

9. To get the overall equation from the given half-reactions, it is necessary to reverse the second half-reaction (which changes the sign of E). Then add the two values.

$$E_{\text{cell}}° = (0.52 - 0.15)\ \text{V} = 0.37\ \text{V}$$

10. The concentrations are nonstandard; therefore, it is necessary to use the Nernst equation.

$$E = 0.37\ \text{V} - \left(\frac{0.0592}{1}\right)\log\frac{[1.50]}{[0.25]^2} = 0.29\ \text{V}$$

11. Inspection of the standard cell potentials shows:

- The reduction of H_2O is easier than the reduction of K^+.

- The oxidation of I^- is easier than the oxidation of water.

$$2\ H_2O(l) + 2\ I^-(aq) \rightarrow H_2(g) + 2\ OH^-(aq) + I_2(s)$$

EXERCISE 18-2

1. $Cr_2O_7^{2-}(aq) + 14\ H^+(aq) + 3\ S^{2-}(aq) \rightarrow 2Cr^{3+}(aq) + 3\ S(s) + 7\ H_2O(l)$

2. $Cl_2(g) + 2\ H_2O(l) + SO_2(g) \rightarrow SO_4^{2-}(aq) + 4\ H^+(aq) + 2\ Cl^-(aq)$

3. $5\ PbO_2(s) + 4\ H^+(aq) + 2\ Mn^{2+}(aq) \rightarrow 5\ Pb^{2+}(aq) + 2\ H_2O(l) + 2\ MnO_4^-(aq)$

4. $2\ Bi^{3+}(aq) + 3\ SnO_2^{2-}(aq) + 6\ OH^-(aq) \rightarrow 3\ SnO_3^{2-}(aq) + 3\ H_2O + 2\ Bi(s)$

5. $4\ MnO_4^-(aq) + 2\ H_2O(l) + 3\ ClO_2^-(aq) \rightarrow 4\ MnO_2(s) + 4\ OH^-(aq) + 3\ ClO_4^-(aq)$

Chemistry of the Elements

EXERCISE 19-1

1. Hydrogen is an exception to many rules.

2. –1 Hydrogen is +1 with most nonmetals.

3. +8 The total of the oxidation numbers for a compound is 0. In this compound, each O has a –2 oxidation number.

4. **Potassium, rubidium, and cesium**   Lithium forms an oxide, and sodium forms a peroxide.

5. hydrogen, H_2

6. gold, Au

7. +5 and –3 Phosphorus is in group 5A, so the highest value precedes the A, and the lowest value is 5 – 8.

8. The ligand serves as the Lewis base.

9. The general equilibrium equation for the formation of a complex is M + x L $\leftrightarrows$ [M L$_x$]

10. Using the answer to question 9, the equilibrium constant expression is products over reactants with the coefficients being exponents.

$$K_f = \frac{[ML_x]}{[M][L]^x}$$

EXERCISE 19-2

1. $Na_2O(s) + H_2O(l) \rightarrow 2\,NaOH(aq)$ Na_2O is a base anhydride.

2. $CaO(s) + H_2O(l) \rightarrow Ca(OH)_2(aq)$ CaO is a base anhydride.

3. $SO_2(g) + H_2O(l) \rightarrow H_2SO_3(aq)$ SO_2 is an acid anhydride.

4. $N_2O_5(s) + H_2O(l) \rightarrow 2\,HNO_3(aq)$ N_2O_5 is an acid anhydride.

5. $NO(g) + H_2O(l) \rightarrow$ no reaction   NO is one of the few nonmetal oxides that is not an acid anhydride.

20
Nuclear Chemistry

EXERCISE 20-1

1. a, b, and d It is important to remember that the mass number is normally not listed.

2. isotopes

3. In each case, there is more than one answer. You may prefer either representation; however, you should be able to recognize them all.

   a. $^4_2\alpha$ or ^4_2He

   b. $^0_{-1}e$ or $^0_{-1}\beta$

   c. γ or $^0_0\gamma$

   d. $^0_{+1}e$ or $^0_{+1}\beta$

4. The following species should go into the blanks.

   a. $^0_{-1}e$

   b. $^0_{-1}\beta$

   c. $^4_2\alpha$ or ^4_2He

   d. $^{252}_{101}Md$

   e. $^{239}_{92}U$

5. **10 days**   Beginning with 100%, one (first) half-life leaves 50%, the next (second) half-life leaves 25%, the next (third) half-life leaves 12.5%, and the next (fourth) half-life leaves 6.25%. Therefore, 40 days equals 4 half-lives.

6. **a.** Nuclear fission is the breakdown of a nucleus into two or more smaller nuclei with the release of energy.   **b.** Fusion is the combining of lighter nuclei into a heavier one.

7. Enter the appropriate values and unit conversions into $E = mc^2$.

$$E = mc^2 = \left(\frac{0.0304 \text{ g}}{\text{mol}}\right)\left(\frac{1 \text{ kg}}{10^3 \text{ g}}\right)\left(\frac{3.00 \times 10^8 \text{ m}}{s^2}\right)^2\left(\frac{1 \text{ J}}{\left(\frac{\text{kg} \cdot \text{m}^2}{s^2}\right)}\right)$$

$$= 2.74 \times 10^{12} \text{ J/mol}$$

8. **25.00%** This is after two half-lives, 21.52 y/10.76 y = 2.

9. **This is after two half-lives, so $t_{1/2}$ = 182 min / 2 = 91.0 min.** One half-life converts 5.00 mg to 2.50 mg and a second half-life converts 2.50 mg to 1.25 mg.

10. It is necessary to use two kinetics equations for first-order reactions: (1) the half-life equation (to find k), and (2) the integrated rate law (to find t). In each case, rearrange the equation and enter the appropriate values.

$$k = (\ln 2)/t_{1/2} = 1.209424 \times 10^{-4} \text{ y}^{-1} \text{ (unrounded)}$$

$$t = -\frac{\ln\left(\dfrac{N_t}{N_0}\right)}{k} = -\frac{\ln\left(\dfrac{58.2\text{ %}}{100.0\text{ %}}\right)}{1.209424 \times 10^{-4} \text{ y}^{-1}} = 4.48 \times 10^3 \text{ y}$$

Organic Chemistry, Biochemistry, and Polymers

EXERCISE 21-1

1. unlimited

2. alkanes, alkenes, alkynes, and aromatic hydrocarbons

3. All types of organic compounds will undergo combustion.

4. The general formula is C_nH_{2n+2}.

5. Both general formulas are C_nH_{2n}.

6. Alkane have the suffix *-ane*.

7. Carbon atoms have four bonds.

8. The most likely site of reaction is at a functional group.

9. This is a polymerization reaction yielding an addition polymer.

$$CH_2{=}CH_2 \ CH_2{=}CH_2 \ CH_2{=}CH_2 \rightarrow {-\!\!-}CH_2{-\!\!-}CH_2{-}CH_2{-}CH_2{-}CH_2{-}CH_2{-\!\!-}$$

10. A polymer of amino acids is a protein (or a polypeptide)

11. All carbon atoms except the top and bottom atoms are chiral.

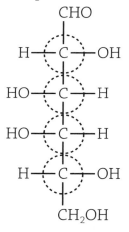

EXERCISE 21-2

1. butane

2. cyclohexane

3. 3-methyl-1-butene

4. 4,4-dimethyl-2-hexyne

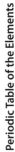

Periodic Table of the Elements

Key:
10	Atomic number
Ne	
Neon	
20.18	Approximate average atomic mass.

1 IA	2 IIA	3 IIIB	4 IVB	5 VB	6 VIB	7 VIIB	8 VIIIB	9 VIIIB	10 VIIIB	11 IB	12 IIB	13 IIIA	14 IVA	15 VA	16 VIA	17 VIIA	18 VIIIA
1 **H** Hydrogen 1.008																	2 **He** Helium 4.003
3 **Li** Lithium 6.941	4 **Be** Beryllium 9.012											5 **B** Boron 10.81	6 **C** Carbon 12.01	7 **N** Nitrogen 14.01	8 **O** Oxygen 16.00	9 **F** Fluorine 19.00	10 **Ne** Neon 20.18
11 **Na** Sodium 22.99	12 **Mg** Magnesium 24.31											13 **Al** Aluminum 26.98	14 **Si** Silicon 28.09	15 **P** Phosphorus 30.97	16 **S** Sulfur 32.07	17 **Cl** Chlorine 35.45	18 **Ar** Argon 39.95
19 **K** Potassium 39.10	20 **Ca** Calcium 40.08	21 **Sc** Scandium 44.96	22 **Ti** Titanium 47.88	23 **V** Vanadium 50.94	24 **Cr** Chromium 52.00	25 **Mn** Manganese 54.94	26 **Fe** Iron 55.85	27 **Co** Cobalt 58.93	28 **Ni** Nickel 58.69	29 **Cu** Copper 63.55	30 **Zn** Zinc 65.39	31 **Ga** Gallium 69.72	32 **Ge** Germanium 72.59	33 **As** Arsenic 74.92	34 **Se** Selenium 78.96	35 **Br** Bromine 79.90	36 **Kr** Krypton 83.80
37 **Rb** Rubidium 85.47	38 **Sr** Strontium 87.62	39 **Y** Yttrium 88.91	40 **Zr** Zirconium 91.22	41 **Nb** Niobium 92.91	42 **Mo** Molybdenum 95.94	43 **Tc** Technetium (98)	44 **Ru** Ruthenium 101.1	45 **Rh** Rhodium 102.9	46 **Pd** Palladium 106.4	47 **Ag** Silver 107.9	48 **Cd** Cadmium 112.4	49 **In** Indium 114.8	50 **Sn** Tin 118.7	51 **Sb** Antimony 121.8	52 **Te** Tellurium 127.6	53 **I** Iodine 126.9	54 **Xe** Xenon 131.3
55 **Cs** Cesium 132.9	56 **Ba** Barium 137.3	57 **La** Lanthanum 138.9	72 **Hf** Hafnium 178.5	73 **Ta** Tantalum 180.9	74 **W** Tungsten 183.9	75 **Re** Rhenium 186.2	76 **Os** Osmium 190.2	77 **Ir** Iridium 192.2	78 **Pt** Platinum 195.1	79 **Au** Gold 197.0	80 **Hg** Mercury 200.6	81 **Tl** Thallium 204.4	82 **Pb** Lead 207.2	83 **Bi** Bismuth 209.0	84 **Po** Polonium (210)	85 **At** Astatine (210)	86 **Rn** Radon (222)
87 **Fr** Francium (223)	88 **Ra** Radium (226)	89 **Ac** Actinium (227)	104 **Rf** Rutherfordium (257)	105 **Db** Dubnium (260)	106 **Sg** Seaborgium (263)	107 **Bh** Bohrium (262)	108 **Hs** Hassium (265)	109 **Mt** Meitnerium (266)	110 **Ds** Darmstadtium (269)	111 **Rg** Roentgenium (272)	112 **Cn** Copernicium	113 **Nh** Nihonium	114 **Fl** Flerovium	115 **Mc** Moscovium	116 **Lv** Livermorium	(117) **Ts** Tennessine	118 **Og** Oganesson

Lanthanide series:
58 **Ce** Cerium 140.1	59 **Pr** Praseodymium 140.9	60 **Nd** Neodymium 144.2	61 **Pm** Promethium (147)	62 **Sm** Samarium 150.4	63 **Eu** Europium 152.0	64 **Gd** Gadolinium 157.3	65 **Tb** Terbium 158.9	66 **Dy** Dysprosium 162.5	67 **Ho** Holmium 164.9	68 **Er** Erbium 167.3	69 **Tm** Thulium 168.9	70 **Yb** Ytterbium 173.0	71 **Lu** Lutetium 175.0

Actinide series:
90 **Th** Thorium 232.0	91 **Pa** Protactinium (231)	92 **U** Uranium 238.0	93 **Np** Neptunium (237)	94 **Pu** Plutonium (242)	95 **Am** Americium (243)	96 **Cm** Curium (247)	97 **Bk** Berkelium (247)	98 **Cf** Californium (249)	99 **Es** Einsteinium (254)	100 **Fm** Fermium (253)	101 **Md** Mendelevium (256)	102 **No** Nobelium (254)	103 **Lr** Lawrencium (257)

Legend:
- Metals
- Metalloids
- Nonmetals

Teacher's Guide

Chemistry is regarded as a "hard" subject because a sizable part of the content is very abstract, math is involved, and the naming of chemical compounds is fairly extensive. Most chemistry teachers and chemical education researchers would agree that simply standing in front of the class and lecturing ("sage on the stage") is not a very effective way to teach chemistry. To learn chemistry, the students must be engaged with the material. It is not a spectator sport—a good deal of practice is involved. There is a lot of similarity between learning chemistry and learning a foreign language. Both require learning a lot of vocabulary. Both have "rules" for putting that vocabulary together. And both require a good deal of practice.

Dealing with Abstract Content

A lot of chemistry can be considered to be abstract, but not as much as many people think. Certainly, atomic structure can be considered to be abstract. We cannot see a proton or electron, but there are ways to have the students have a hands-on experience. The lab portion of chemistry is crucial in providing the students a hands-on experience with "difficult" or abstract concepts. Hands-on is ideal, but demonstrations are good also (especially if you have a limited amount of equipment and/or budget). Use a lot of analogies and examples, and relate the concept to everyday experiences that the student might have experienced.

Dealing with Math

Yes, there is a significant amount of math in chemistry. But at the high school level (as well as most introductory chemistry classes in college), the math is mostly arithmetic and a little algebra (solving for x). Teaching the students to use the method of Unit Analysis (sometimes called the Factor Label Method) can allow them to be successful while developing chemical intuition. See Chapter 1 for a description of this method and practice problems. And the last step of the Unit Analysis method is to make sure they have generated a reasonable answer. For example: Question: How many molecules of water are in 18.0 grams of water? Common Student Answer: 1 molecule. "Incorrect! 18.0 grams of water is about 18 mL, a nice sip of water, but you are saying that there is only one molecule there! It must be a large one! You have confused molecules and moles." If the student had simply thought about whether the answer was reasonable, they could have avoided an incorrect answer.

Another good technique in problem solving is to generate the setup for the problem and estimate the answer. This is a great way of checking your answer or quickly eliminating one or more answers in a multiple-choice question.

Dealing with Chemical Nomenclature

Naming chemical compounds correctly is an important concept in chemistry. In my introductory chemistry classes, I include chemical nomenclature on every quiz and every exam. I start off by having the students learn the common monovalent ions and figure out the formulas, and then have them progress to the polyvalent ions and name covalent compounds. Flashcards really help here.

How (and When) to Use This Book

The easy answer is to use it when you think it will help your students. This book has just about all the topics that are in our general chemistry college textbook (*General Chemistry: Understanding Moles, Bonds, and Equilibria, Volumes 1 and 2*), even if we don't always go as deep into the content as in our college text. We designed this text to supplement any high school textbook, but you certainly could use it as a stand-alone text for Chem 1 or Chem 2. It would also work well as a review of chemistry before an AP Chemistry course. (We do have a study guide specifically for AP Chemistry: *5 Steps to a 5 AP Chemistry*.) It could even be used as a test review. The sky is the limit.

Additional Pedagogical Resources

Teaching chemistry is just as much about how to teach as what to teach. The following resources can provide valuable content and insights to effective pedagogy.

The **American Association of Chemistry Teachers** (AACT) (teachchemistry.org) is an organization that is geared toward high school chemistry teachers. It publishes classroom resources, professional development, news, and so on. You get a lot of useful information for your dues.

The **American Chemical Society** (acs.org) is the premier organization devoted to chemistry. It has a division that is devoted to chemical education (divched.org). The ACS publishes *ChemMatters* magazine four times a year. It is devoted to clarifying chemistry for high school students and helping them see the connections between chemistry and everyday life. If you join AACT you receive a complimentary subscription to *ChemMatters* and have access to the *ChemMatters* archive.

Another professional organization that deserves consideration is the **National Science Teaching Association** (NSTA, nsta.org). NSTA has a great deal of resources available for their members, including lesson plans, books, and so on. This organization covers not only chemistry but also all the other sciences. This is invaluable for chemistry teachers who are also teaching physics, biology, and the like.

Flinn Scientific (flinnsci.com) is a wonderful source for chemical supplies and chemicals. They have a real devotion for educating teachers in how to conduct safe science at the public school level. Their hard copy catalog contains a wealth of information on chemical disposal and the safe handling of chemicals. It makes great bedtime reading!

In response to COVID-19 and the need to provide virtual instruction, many companies developed classroom instructional materials. These materials are also valuable in the face-to-face classroom. A simple Google search will generate dozens of useful hits. It is easy to get overwhelmed with the number of hits, so we suggest you specify the topic to narrow it down (i.e., teaching kinetics, teaching atomic structure, and so on).